Biological Systems: Complexity and Artificial Life

Authored by

Jacques Ricard

Honorary Director of the Jacques Monod Institute
CNRS
Universities Paris 6 and Paris 7
Paris cedex 5
France

CONTENTS

PREFACE

For many years biology has been rooted in a reductionist attitude probably derived from Descartes philosophy that consists in splitting the problems into simple elements that should help solve these problems. These elements are specific macromolecules, namely nucleic acids and proteins. For many years it was, explicitly or implicitly, postulated that major biological problems could be solved from the knowledge of structure and function of nucleic acids and proteins. The term "molecular biology" clearly implied that it is possible to reduce *biological problems* to *molecular entities*. In fact many difficulties arise when one attempts at reducing biological *problems* to *biomolecules*. The main reason for this difficulty is that any living organism is a *complex system*.

The concept of complex system has played a major role in contemporary science, particularly in biology and social sciences. Complex systems display a number of features that are stated below [1,2].

> *Complex systems are made up of a large number of elements in interaction.*
>
> *Such systems are not disordered but display a certain degree of order and organization.*
>
> *These systems display nonlinear effects and feedback loops.*
>
> *From a thermodynamic viewpoint, a complex system usually operates away from thermodynamic equilibrium and requires both matter and energy to maintain its organization.*
>
> *A complex system is in a dynamic state and possesses a history. Its present state is determined by its past behaviour.*
>
> *The interactions between the elements of the system take place over a short range and are therefore local. These features and others have already been mentioned earlier [1].*

It is therefore possible to point out the differences between molecular and systems biology. As briefly pointed out above, molecular biology attempts at understanding properties of a living organism from the individual properties of some type of biomolecules such as DNA, RNA and proteins. Systems biology explains the global properties of the living organisms from the dynamic properties of *systems* of biomolecules. But a system, a network, of biomolecules can be more than a simple set of macromolecules. It involves multiple interactions between its elements and display the *emergence* of novel properties, that is properties that are "novel" relative to the individual properties of the elements of the system. Put in other words, such systems display *emergence* of properties that are not borne by any individual element of the global system. It is evident then that the *global* properties of these systems can only be studied with the help of mathematics which then appears an unavoidable aspect of systems biology. In this line, it becomes possible to approach some aspects of biological problems through the use of simple mechanistic models of living systems and this approach can be considered some kind of *systems biology.*

Many biological problems can now be studied trough the concept of system. Two books have been devoted to these problems [2, 3] and it is therefore useful, to use some of these problems as a material to approach this field of *complexity* and *systems biology*.

ACKNOWLEDGEMENTS

The author wishes to thank his wife Käty for her constant support as well as Robin Coombes for carefully reading his manuscript.

CONFLICT OF INTEREST

The author confirm that, as far as he knows, the book content does not posses any conflict of interest.

REFERENCES

[1]   Cilliers, P. Complexity and Postmodernism (1998), Routledge, London and New York.
[2]   Ricard, J. Biological Complexity and the Dynamics of Life Processes.(1999), Elsevier, Amsterdam, Lausanne, New York.
[3]   Klipp, E, Liebermeister, W, Wierling, C, Kowald, A. Lehrach, H. and Herwig, R.(2010), Systems Biology, Wiley-Blackwell.

Jacques Ricard

Honorary Director of the Jacques Monod Institute
CNRS, Paris cedex 5,
France
E-mail: Jkricard@aol.com

2

Send Orders for Reprints to reprints@benthamscience.net

CHAPTER 1

The Vision of Classical Molecular Biology and its Limits

Abstract: Molecular approach has been, for decades, the classical way for studying biological problems. It implies that the essential biological properties can be reduced to the study of two classes of biomolecules, namely nucleic acids and proteins. The interest and the limits of the so-called molecular biology are discussed in this Chapter.

Keywords: Reduction, Reductionism, Ontological reductionism, Descartes, High-level theory, Low-level theory, Molecular biology, Eukaryotic cells, Axiom, Ligand.

Molecular biology emerged around 1950 from strong connections between genetics and biochemistry. Its aim is to understand the essence and mechanisms of biological processes through the knowledge of the nature and function of biological macromolecules. As a matter of fact, molecular biology referred mostly to two classes of macromolecules, nucleic acids and proteins. There is a simple and logical reason for such a situation as the so-called genetic message appeared coded in the structure of nucleic acids, DNA (deoxyribonucleic acid) and RNA (ribonucleic acid), before being expressed under the form of proteins. In this classical vision of biological processes, the main features of living systems should be stored in the DNA molecules and then expressed as proteins. In this perspective, one could conceive that the knowledge of the genome and the proteome could be sufficient to predict and explain the morphology, the physiology and even some aspects of the pathology of living organisms. This reasoning is obviously reductionist in its essence for it postulates that it is possible to understand biology of living systems through a study of biosynthesis, structure and function of two classes of macromolecules, namely nucleic acids and proteins.

1- REDUCTION

In its classical aspects, molecular biology is based upon the philosophical concept of reduction that dates back to Descartes' fundamental books "Règles pour la direction de l'esprit" [1] and "Discours de la méthode" [2]. In its ontological formulation, reductionism is based upon three fundamental principles:

The search for simplicity behind the apparent complexity of phenomena;

The attempt to reduce complex processes to simple phenomena;

The deep conviction is based on the understanding that the world relies upon a linear sequence of causes and effects.

Basically, reductionism is defined by the process of thought that aims at formulate the principles and results of a theory, T_h in terms of principles and results, C_l, of another theory, T_l more general and embracing than T_h. T_h is called a high-level theory and T_l a low-level theory [3-12]. In order to allow such reduction, concepts and predicates, C_h, of the high-level theory (T_h) should be present among the concepts and predicates, C_l, of the low-level theory (T_l). Hence it follows that

$$C_h \subset C_l \tag{1.1}$$

In order for this relation to be meaningful, concepts and predicates of the high-level theory should also be concepts and predicates of the low-level theory. If any high-level theory could be reduced to a low-level one, this would imply that Science is unique and that division in several Sciences is just an illusion. Insofar as biological processes can be reduced to physical ones, this could be expressed as

$$C_{biol} \subset C_{phys} \tag{1.2}$$

where C_{phys} and C_{biol} represent concepts and predicates of physical and biological theories, respectively. As biology is more complex than physics, relation (1.2) above is equivalent to

$$C_{comp} \subset C_{simple} \tag{1.3}$$

where C_{comp} and C_{simp} represent concepts and predicates of complex and simple systems, respectively. In this perspective, understanding a complex system would be equivalent to understand the nature and relations of their constituents.

Reduction can be conceived at two different levels: the methodological and ontological levels. As far as the first level is concerned, there is little doubt that one should know the properties of the elements of the system before trying to understand the properties of the system itself. Ontological reduction, however, is a completely different problem. It implies that no unexpected property can emerge out of the interactions that exist between the elements of the system. In this perspective, all the properties of the system should already be present in the constitutive elements of the system. The term "molecular biology" precisely means that the biological properties of a living organism are borne by their macromolecules. If such a situation were valid, this would imply that any property of a living organism were borne by material entities *i.e.* macromolecules. No emergence of a property could ever be generated thanks to the interactions between different elements of the system. According to Monod [13] for instance « l'organisation d'ensemble d'un édifice moléculaire était contenu en puissance dans la structure de ses constituants, mais ne se révèle, ne devient actuelle, que par leur assemblage » or « l'information était présente, mais inexprimée dans ses constituants…. La construction épigénétique d'une structure n'est pas une création, c'est une révélation ». If this perspective were correct, living organisms could not be considered as systems. As we shall see, it is clear that the ideas that are developed in this book do not conform to these views. Monod's statement above has two implications: the first one is that the properties of an enzyme should be the same if considered in isolation or within a living cell; the second one is that the conformation of the enzyme, and therefore its functional properties, is independent upon the fact that a ligand could bind to the enzyme and modify its conformation. In other words the enzyme properties should be the result of interactions between the protein and its environment. It is of interest to discuss briefly these two points.

Most enzymes in eukaryotic cells are not free but associated with different cell organelles such as membranes and cell walls. As a result of these interactions, the properties of the enzyme could be changed in such a way that the behavior of the free enzyme in solution could have little to do with the one occurring under the form of a multi-enzyme complex, free or bound to a membrane. Another striking effect, which will be discussed in a forthcoming Chapter, is that these organelles are carrying positive, or negative, charges in such a way that the properties of these bound enzymes are qualitatively changed and display few similarities with those of the same enzyme in free solution. It then appears difficult to conclude that the properties of an enzyme in free solution could be identical to those of the same enzyme in the living cell.

As previously mentioned, another axiom inherent to the principle of ontological reductionism is that the conformation of an enzyme, and therefore its properties, are independent upon the interactions they may possess with different ligands. Many biochemical studies have shown that different ligands can bind to proteins and result in a conformation change associated with a change of their activity. In the general frame of the implicit postulates of molecular biology one can assume that the protein pre-exist under two or several conformations that can bind ligands with a different affinity. In this perspective, the ligand does not *induce* a novel protein conformation that possesses novel properties but rather *selects* an already existing conformation with different properties. This is what has been postulated by some founding fathers of molecular biology in the frame of the so-called *allosteric model* [14]. In such a model, the same protein pre-exists under different conformations in equilibrium and the ligand selects one of these conformations. But the same results may be explained in a different model in which the protein exists under one conformation state only [15]. It is the binding of the ligand to the protein that induces the *emergence* of a new conformation of the protein. In this model, there is an *instruction* given by the medium to the protein to adopt a new conformation. Hence the properties of the protein would result from the interaction between the protein and its environment. According to molecular biologists the first model would be "*selectionist*" and its properties would be solely defined by its own structure only. Alternatively, the second model would imply the existence of an *instruction* called *induced-fit* given by the medium to the protein. For that reason, the corresponding system would be considered "*instructionist*". The first "*selectionist*" model was favored by molecular biologists because it was considered "*darwinian*" in its essence, whereas the second "*insructionist*" model was rejected by the same biologists as being "*lamarkian*".

It is clear that this controversy is supported neither by theoretical considerations nor by the experimental results. In the case of the so-called allosteric model, it is not true that the ligand *selects* some binding sites on the protein molecule for if the corresponding concentration is high enough it will bind to *all* the available sites. Even if there is no site selection, there is selection of a conformation.

2- FOUNDATIONS OF MOLECULAR BIOLOGY

As we have seen, classical molecular biology aims at understanding biological processes from the knowledge of the structure, the function and the biosynthesis of nucleic acids and proteins. As previously mentioned, this idea appeared promising and logical for the "genetic message" is coded and stored as deoxyribonucleic acid, then expressed under the form of proteins. Hence, in this classical vision of life, the main trait of any living system would be stored in DNA structure, and expressed as proteins. If these statements were valid they would justify the reductionist vision of biological systems.

2.1- DNA and Heredity

The reductionist version of biology is rooted in the discovery, in 1953, of the three-dimensional structure of DNA [16, 17]. This remarkable achievement was made possible only because a previous work by Avery, McLeod and McCarthy [18] had shown that a chemical substance, DNA, extracted from a virulent strain of *Pneumococcus* could induce the transformation of a non-virulent strain of *Pneumococcus* into a virulent one. This was the demonstration that a biological character, virulence, could be carried by a DNA molecule. The work of Watson and Crick [16, 17] had the merit to make evident how a DNA molecule could replicate. As everyone knows today, this molecule is made up of four different nucleotides, adenine, thymine, cytosine and guanine that form a pair of helices associated through hydrogen bonds between adenine and thymine as well as cytosine and guanine. This kind of structure allows understanding how a DNA molecule can replicate. Breaking the hydrogen bonds generates two single strands that can associate, through hydrogen bonds, free, unbound, complementary nucleotides. The same process made also evident that the same DNA molecule could be used to generate a single-stranded molecule, RNA. This can be achieved by associating, through hydrogen bonds, adenine to uracile, instead of thymine. One obtains in that way a single-chain macromolecule, called messenger RNA (mRNA) in which the sequence of motifs is the same as that of the corresponding DNA. This is the process of transcription of DNA into a single-stranded molecule, mRNA. Owing to the double helix structure of the DNA, one could explain two different functions of genes, namely the ability to reproduce with the same structure and give birth to different molecules that keep the same sequence of different motifs. Hence it appears that biological functions of genes can be *reduced* to the structure of the DNA molecule and of its transcripts, mRNAs.

The single-strand mRNAs can in turn be expressed as proteins. In other words, the DNA message can be translated into a protein message. But this process necessitates a code, the so-called genetic code, which explains the sequence of aminoacids in proteins. As a matter of fact, DNA can be conceived as a message written in a four-letter alphabet (adenine....thymine, thymine....adenine, cytosine....guanine and guanine....cytosine). The DNA message is transcribed in a mRNA message written in a four-letter alphabet (adenine....uracile, uracile....adenine, cytosine....guanine and guanine....cytosine). Last but not least, the mRNA message is translated into a protein message. Whereas the DNA and mRNA messages are the same but written in alphabets that possess four letters, the mRNA and protein messages are obviously the same but written in alphabets possessing a different number of letters (four letters for the mRNA message and 20 letters for the protein message). The last point is important for it shows that the genetic code should involve $4^3 = 64$ protein codons to specify the 20 different aminoacids present in proteins. This explains that the code is degenerate and that some codons are nonsense.

2.2- Nature of Genes

Initially, around year 1950, a gene was identified with a DNA segment expressed as a specific protein. However, it was soon discovered that gene expression requires a non-coding segment located upstream the coding sequence. This finding did not change much the prevailing ideas about the nature of genes. The situation became even more complex when it appeared that eukaryotes, *i.e.* organisms possessing a well differentiated nucleus, display a non-coding regulatory sequence controlling several genes located in the genome far away from the genes they control. Moreover in these eukaryotic organisms the same coding sequence called exon is interrupted by several non-coding sequences, the introns. As a consequence of this organisation, only the exons are expressed as mRNAs. Moreover the connexion between the exons could be effected in several different manners in such a way that one gene controls several different mRNAs that can be expressed as different proteins. Hence the dogma one gene→one protein is violated. The same kind of result can be obtained owing to a frame-shift taking place during the process of mRNA translation. It then appears that the same gene gives birth to different proteins and, for that reason, can be considered a fuzzy concept.

2.3- Genomes

In order to avoid the above mentioned difficulty in defining the concept of gene in eukaryotic organisms it may seem easier to study the global properties of sets of genes, *i.e.* the properties of *genomes*. From this point of view, molecular genetics gives birth to *genomics*. A good example of this type of approach is given by a celebrated research program called the *human genome project*. Its ambition is to gain new information about biological complexity by identifying and studying all the genes present in human species. The ambition of this project is highly reductionist in its essence for his leaders explicitly, or implicitly, assumed that if we know all the genes of the human species we should know nearly everything about man and woman. This is clearly a reductionist ambition. Whatever that may be, if it were possible to deduce all the basic properties of a living organism from its genes, one should expect a strong correlation to occur between the number of genes of a living species and its position in the *scala naturae*. What the *Human Genome Project* has demonstrated is the lack of correlation between the number of genes of organisms and their degree of sophistication. On the basis of classical comparative biology eukaryotes are more complex and sophisticated than prokaryotes and the same situation occurs for animals as compared to plants. However if the number of genes of the primitive worm *Caenorhabditis elegans* is compared to that of *Drosophila melanogaster* who is more complex and sophisticated, the result obtained is the converse than the one expected. *Caenorhabditis* has 18424 genes and *Drosophila* 13600. Similarly the tiny plant *Arabidopsis thaliana* has 25500 genes and man only 35000! There is no correlation between the number of genes of animals or plants and the *scala naturae*.

An interesting result brought by a number of studies is that sets of genes act in a coordinated manner. This conclusion was already present in the seminal work of

Jacob and Monod [19]. It has been possible to associate different genes as to form artificial operons. If, in such a system, a gene is activated by the product of another gene the resulting system can display several steady states and therefore several functional activities. These results are hardly compatible with the idea that the functional activity of a gene battery can be understood by reducing its global activity to the individual activities of a population of unconnected genes.

3- CONTROL OF DIFFERENTIATION

What molecular genetics has shown is that several DNA regions, called genes, are able to control synthesis of specific proteins. The main question that arises is to understand how the specific synthesis of RNAs and proteins give rise to non-uniform spatial organization of the embryo. In a pluricellular organism, all the cells of the young embryo originate from the successive divisions of a unique fertilized egg. Hence all these cells should possess the same genes. Yet depending on their location in the young embryo, the cell differenciate in different manners. It is therefore obvious that morphogenesis is a problem that cannot be reduced to the action of isolated molecules.

A first approach to this problem was offered by Child [20] who supposed that there exists in young embryos gradients of "morphogens". A theoretical model was developed by Turing [21] who showed that, at least, three conditions are required to generate different evolution of the cell, depending on their location in the embryo:

-   Gradients of at least two substances should appear in the initially homogeneous medium.

-   The existence of these morphogenetic gradients implies that the medium is away from thermodynamic equilibrium.

-   One of these morphogens stimulates its own formation through an autcatalytic process whereas the other morphogen inhibits the action of the first one.

Gradient of morphogens have been discovered in Drosophila eggs and in young embryos [22, 23]. The so-called Drosophila follicle possesses sixteen cells. One of them is the oocyte, the others are called the nurse cells. The young embryo is formed from the division of the oocyte. Proteins and mRNAs are transferred from the nurse cells to the oocyte. Hence Drosophila follicle appears as a non-equilibrium system displaying both gradients of proteins and mRNAs that can activate or inhibit various genes. In fact, strictly speaking, there is no gene responsible for the emergence of the form of the embryo but only *systems* involved in the transport, under non-equilibrium conditions, of molecules that possess a role in the organization of the embryo. It is the location, in space and time, of these molecules, their gradients, that play the major role in porphogenesis.

4- GENES, DISEASES AND ENVIRONMENT

An idea that has been popular among molecular biologists is what has been called by Rees a "genocentric" vision of diseases.[1] According to this view, many diseases would be caused by mutations. In this optic, recovery would imply some sort of "correction" of genes through genetic engineering. This would be particularly true for cancers and cardiovascular diseases. As pointed out by Strohman, it appears that at least 80% of these diseases are due to environmental effects and, in particular, to what people eat. Two facts seem to confirm this conclusion. The supervening of cancers in a population dramatically changes if the diet of this population changes. For instance, the frequency of cancer in Japan is much lower than in the USA. However the frequency of this disease becomes about the same for American and Japanese populations living in the USA.

It is striking to note that the ontological reductionist approach of biological problems has been so far of little use in the case of AIDS. As a matter of fact the virus involved in this disease has been isolated and its genome deciphered, but these achievements have been of little use to cure this disease.

5- FROM GENOTYPE TO PHENOTYPE

Understanding and solving biological problems implies that one can understand the logic of four transitions.[2] The first one is a genome → transcriptome transition

---

[1]Rees, J. (2002) Science 296, 695-698.
[2] (1) Strohman, R. (2002) Science 296, 701-703.

which implies that gene products, namely mRNAs, can undergo specific alterations. The second one is the transcriptome→proteome transition that involves post-translational modifications of enzyme structure. The last one is the proteome→dynamic systems transition. It is this transition that is of particular interest for a dynamic system cannot be understood if one solely refers to the principles of ontological reductionism. There is no doubt that an isolated enzyme does not behave the same way in a test tube and in a living cell. There are at least three reasons for that. The first one is the fact that an enzyme in the cell is connected to many others as to form a network, in such a way that its apparent properties are dependent upon many other enzyme reactions. The second reason, which is a particular case of the previous situation, is that the enzyme is involved in a "cascade" process leading to unexpected events. The third reason is an interaction of the enzyme and its substrate with fixed charges and cell organelles, such as cell walls. All these effects are systemic and cannot be understood on the sole basis of the reductionist approach. The last transition is a transition dynamic systems→phenotype. This transition takes into account how the other transitions are involved into the making up of the phenotype.

6- GENERAL CONCLUSIONS

The discovery of the DNA structure, which offered an apparently simple explanation of gene reproduction and expression, convinced a number of scientists that the "secret of life" had been discovered. Reductionism, which was initially methodological, became soon ontological, and a number of scientists became, implicitly or explicitly, convinced that it is sufficient to know the parts, the elements, of a system to understand the behavior of the system itself. This kind of ontological reductionism, which dates back to Descartes, is associated with the process of reification of biological functions. According to this view, any macroscopic biological function should be physically borne by some molecular entities and the aim of molecular biology is precisely to discover the nature of these entities.

An alternative of these views is to consider living organisms as systems that display both complexity and emergent properties. As it will be shown in this book the concept of emergence, which has often been considered a metaphysical one, can be given a physical meaning and implies that some elements of a system can interact as to generate novel and unexpected effects. Systems, whatever their nature, can often be subjected to mathematical modelling that can explain experimental results and allow discovering new ones. The aim of this book is precisely to offer a brief account of this new science of systems biology and biological complexity.

REFERENCES

[1]   Descartes, R. (1959) rééd. Règles pour la Direction de l'esprit. Troisième édition, traduction et notes de J. Sirven, Vrin, Paris.
[2]   Descartes, R. (1992) rééd. Discours de la Méthode. Flammarion, Paris.
[3]   Hempel, C. G. (1970) Philosophy of Natural Science. Engelwood Cliffs, New Jersey, Prentice Hall.
[4]   Hull, D. L. (1988) Science as a Process. University of Chicago Press, Chicago.
[5]   Nagel, E. (1961) The Structure of Science. Harcourt, Bruce and World, New York.
[6]   Robinson, J. D. (1986) Reduction, explanation and the quests of biological research. Philosophy of Science, 53, 333-353.
[7]   Ayala, F.J. and Dobzhansky, T. eds. (1974) Studies in the Philosophy of Biology. Macmillan, London.
[8]   Bock, G.R. and Good, J. A. eds. (1988) The limits of Reductionism in Biology. Novartis Symposium Foundation, John Wiley and Sons, Chichester, New York.
[9]   Thorpe, W. H. (1974) Reductionism in Biology. In Ayala F. J. and Dobzansky T. eds. Studies in Philosophy of Biology. Macmillan, London, pp. 109-138.
[10]   Beckner, M. (1974) Reduction, hierarchies and organization. In Ayala F. J. and Dobzansky T. eds. Studies in the Philosophy of Biology. Macmillan, London, pp. 163-177.
[11]   Ricard, J. (2001) Complexity, reductionism and the unity of Science. In Agazzi E. and Faye J. eds. The Problems of the Unity of Science. World Scientific, New Jersey, London pp. 97-105.
[12]   Ricard J. (2001) Reduction, integration, emergence and complexity in biological networks. In Agazzi E and Montecucco L. eds. Complexity and Emergence, World Scientific. New Jersey, London pp. 101-112.
[13]   Monod (1970) Le Hasard et la Nécessité. Editions du Seuil, Paris.
[14]   Monod J., Wyman, J. and Changeux, J. P. (1965) On the nature of allosteric transitions. J. Mol. Biol. 12, 88-118.
[15]   Koshland, D. E., Nemethy, G. and Filmer, D. (1966) Comparison of experimental binding data and theoretical models in proteins containing subunits. Biochemistry 5, 365-385.
[16]   Watson, J. D. and Crick, F.H.C. (1953a) Molecular structure of nucleic acid. A structure for deoxyribose nucleic acid. Nature 171, 737 -738.
[17]   Watson, J. D. and Crick, F. H. C. (1953b) Genetic implications of the structure of deoxyribonucleic acid. Nature 171, 964 – 967.
[18]   Avery, O. T., Mac Leod, C. M. and Mac Carty, M. (1944) Studies on the chemical nature of the substance inducing transformation of pneumococcal types. Induction of transformation

by a deoxyribonucleic acid fraction isolated from Pneumococcus Type III. J. Exp. Med. 79, 137-158.

[19]   Jacob, F. and Monod, J. (1961) Genetic regulatory mechanisms in the synthesis of proteins. J. Mol. Biol. 3, 318-356.

[20]   Child, C. M. (1941) Patterns and Problems of Development. University of Chicago Press, Chicago.

[21]   Turing, A. M. (1952) The chemical basis of morphogenesis. Philos. Trans. Royal Soc. London B 237, 37 – 72.

[22]   Driever, W. and Nusslein-Volhard, C. (1988) A gradient of bicoid protein in *Drosophila* embryos. Cell, 54, 83 – 93.

[23]   Driever, W. and Nusslein-Volhard, C. (1988) The bicoid protein determines position in the *Drosophila* embryo in a concentration dependant manner. Cell, 54, 95-1.

Send Orders for Reprints to reprints@benthamscience.net
Biological Systems: Complexity and Artificial Life, 2014, 15-34

CHAPTER 2

Biological Systems, Identity, Organization and Communication

Abstract: A living organism is a complex system that possesses an identity and emergent properties. Moreover such a system is able to communicate with its neighbours and cannot violate physical laws, for instance physical laws involved in communication of messages.

Keywords: Identity of a system, Information of a system, Information associated with an event, Shannon entropies, Source of information, Destination of information, Conditional information, Central dogma of molecular biology, Genetic code, Protein lattice, Shannon communication theory, Emergence of information, Information and spatial organization of a system.

A living cell cannot be considered a mere reservoir where many enzyme-catalyzed reactions take place. It is a highly complex system, some kind of network that possesses an identity that can, more or less, be reproduced from generation to generation. Molecular biology has taught us that, in present days, the information of living organisms is stored in DNA, transcribed as a sequence of RNA bases and finally expressed as aminoacids. Moreover any living organism is a system of connected elements in such a way that the global properties of living systems result from the interactions between their constitutive elements. In a way, most of the global properties of this system can be considered *emergent properties*. Moreover many communication processes take place in living organisms. Even though the most celebrated of these communication processes is the information transfer between DNA and proteins it is far from being the only one....Last but not least, living organisms are spontaneously able to evolve. There is some kind of progress in evolution for this process is exerted through the selection of advantageous mutations, but there is also little doubt that some kind of self-organization takes place during the evolution of living systems.

1– IDENTITY AND INFORMATION OF THE ELEMENTS OF A SYSTEM

The concept of identity of a material element belonging to a system dates back to Aristotle. It can be considered both an ontological principle that gives a material

Jacques Ricard

entity its essence and the ability we have to identify this material entity among many others. This ontological principle, as well as the ability we have to identify material entities, can be defined as information. From the definition of the concept of information emerges a mathematical formulation of this concept. As a matter of fact, it appears evident that the definition of information of a material entity should be related to its probability of occurrence. In particular, the larger the probability of occurrence of an event and the smaller the information brought about by this event.

If an event, x_i, has a probability of occurrence $p(x_i)$, the information associated with the supervening of this event is

$$h(x_i) = f\left\{\frac{1}{p(x_i)}\right\} \tag{2.1}$$

where f is an increasing function. In order to determine the nature of this function, one can take account of an event that depends upon two other independent events x_i and y_j. It is then evident that the corresponding function, $h(x_i, y_j)$, will be equal to the sum of $h(x_i)$ and $h(y_j)$. One has then

$$h(x_i, y_j) = h(x_i) + h(y_j) \tag{2.2}$$

It follows from this reasoning that the simplest h function that meets this requirement is a logarithmic function for one has then,

$$h(x_i, y_j) = -\log p(x_i, y_j) = -\log p(x_i) - \log p(y_j) \tag{2.3}$$

If now the events x_i and y_j are not independent the well-known Bayes theorem requires that,

$$p(x_i, y_j) = p(x_i)p(y_j|x_i) = p(y_j)p(x_i|y_j) \tag{2.4}$$

where $p(y_j|x_i)$ and $p(x_i|y_j)$ are the conditional probabilities of occurrence of y_j and x_i given that x_i and y_j have already occurred. It follows from this relationship that

$$h(x_i, y_j) = h(x_i) + h(y_j | x_i) = h(y_j) + h(x_i | y_j) \tag{2.5}$$

It appears from this expression that the interaction between x_i and y_j may generate an increase, or a decrease, of the corresponding information. In order to express quantitatively the amount of information generated, or consumed, through the interaction between x_i and y_j one can define a new function, $i(x_i : y_j)$, as

$$i(x_i : y_j) = h(x_i) + h(y_j) - h(x_i, y_j) \tag{2.6}$$

taking advantage of expressions (2.5), expression (2.6) can be rewritten as

$$i(x_i : y_j) = h(x_i) - h(x_i | y_j) = h(y_j) - h(y_j | x_i) \tag{2.7}$$

if

$$h(x_i) > h(x_i | y_j) \text{ and } h(y_j) > h(y_j | x_i) \tag{2.8}$$

the interaction between x_i and y_j produces consumption, and possibly migration, of information. If alternatively,

$$h(x_i) < h(x_i | y_j) \text{ and } h(y_j) < h(y_j | x_i) \tag{2.9}$$

information is generated through the interaction between x_i and y_j. In the first case the system is defined as *integrated* and in the second case it is considered *emergent*. As we shall see latter, this scientific concept of *emergence* is essential in the sciences of complexity.

In the above reasoning it was implicitly assumed that only two entities, x_i and y_j interact. This is not compulsory, however, and it is quite likely that many entities can interact. This matter will be discussed latter in Chapter 3.

2– THE PRINCIPLES OF SHANNON COMMUNICATION THEORY

Relationships (2.5)-(2.7) are reminiscent of the so-called Shannon entropies of the communication theory which has been intuitively used to explain how genes control protein synthesis. Hence it is important to recall some of the main ideas of Shannon's communication theory as applied to biological systems [1-7].

The main feature of a communication channel is to be able to convey a message, generated by a source, through a channel up to a destination. The message is encoded in a sequence of symbols that can be considered an alphabet. During the communication process, random alterations of the message usually take place. These errors are the noise of the system. The message, altered by the noise, is decoded, then encoded into a new alphabet and, finally, transferred to its destination. Hence, from a mathematical viewpoint, the code is the mapping of the letters of an alphabet of a probability space Ω_X onto the letters of another alphabet of a different probability space Ω_Y. This is precisely what is taking place during the information transfer from DNA to proteins. The DNA message is expressed first as mRNA, *viz.* the same message written in a four-letters alphabet, namely adenine-thymine, thymine-adenine, cytosine-guanine and guanine-cytosine. This message is then rewritten in a different four-letters alphabet involving adenine, uracil, guanine and cytosine. This molecule is a mRNA that specifically associates another RNAs called tRNAs. This binding process involves three pairs of complementary bases of both mRNA and tRNA. Hence we have a coding system in which three bases (a codon) are required to specify an aminoacid. The classical scheme for transfer and decoding is shown in Fig. **1**.

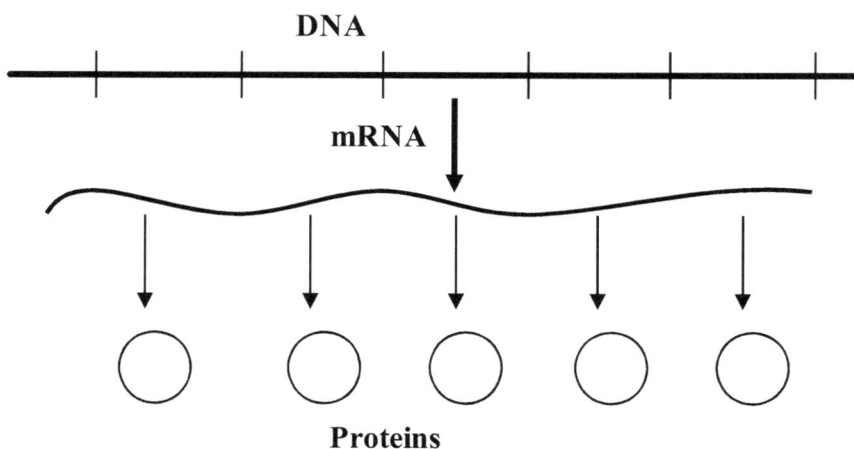

Figure 1: Schematic representation of information transfer from DNA to proteins.

The concept of information has already been defined by the function h (equation 2.1). If we have two independent events pertaining to different alphabets, the communication process, in such an ideal system, implies that

$$h(x_i, y_j) = h(x_i) + h(y_j) \tag{2.10}$$

and one can define the corresponding Shannon entropies

$$H(X) = \sum_i p(x_i)h(x_i) = -\sum_i p(x_i) \log p(x_i) \tag{2.11}$$

$$H(Y) = \sum_j p(y_j)h(y_j) = -\sum_j p(y_j) \log p(y_j)$$

In conventional Shannon theory, the alphabet comprises only values 0 and 1 in such a way that one has to use logarithms of base two. This point will not be discussed any further here. If relationship (2.1) holds then one has

$$H(X,Y) = H(X) + H(Y) \tag{2.12}$$

If, however, the two variables are not independent, which means that the entropy of the source depends upon that of the destination or, conversely, that the entropy of the destination depends upon that of the source, one has

$$H(X,Y) = H(X) + H(Y|X) = H(Y) + H(X|Y) \tag{2.13}$$

where

$$H(X|Y) = \sum_i \sum_j p(x_i, y_j)h(x|y) \tag{2.14}$$

$$H(Y|X) = \sum_i \sum_j p(x_i, y_j)h(y|x)$$

In the case of the communication of a message, $H(X|Y)$ represents the entropy of the source given the entropy of the destination. Alternatively, $H(Y|X)$ is the entropy of the destination given that of the source. The information $I(X:Y)$, which is *effectively* transferred from the source to the destination, is

$$I(X:Y) = H(X) - H(X|Y) = H(Y) - H(Y|X) \tag{2.15}$$

Hence a communication channel is characterized by the mutual information, $I(X:Y)$ associated with a reversible transfer between X and Y and to a noise $H(X|Y) + H(Y|X)$.

A typical information channel is shown in Fig. **2**. It is important to stress that

$$H(X) > H(X|Y) \text{ and } H(Y) > H(Y|X) \tag{2.16}$$

in such a way there cannot exist any process of emergence of information in a communication channel. This is a situation completely different from that described earlier in expression (2. 9)

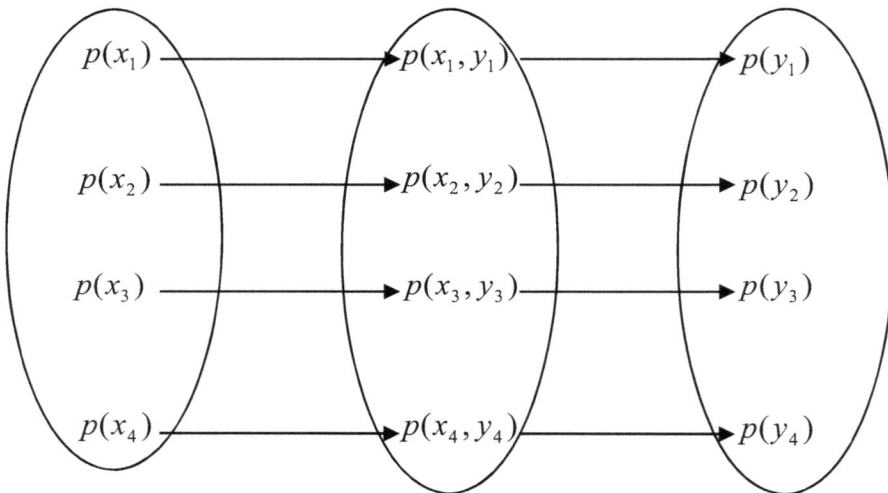

Figure 2: Mathematical description of the information transfer from Ω_X to Ω_Y.

In fact the information transfer between Ω_X and Ω_Y is reversible if the two probability spaces are isomorphic (Fig. **2**) which implies that the numbers of letters of the two alphabets are the same. If, on the other hand, the two probability spaces do not contain the same number of elements, the communication process becomes unidirectional. A good illustration of this situation is offered by the genetic code. As previously mentioned, the genetic message is written in the DNA structure, using a four-letter alphabet, namely adenine-thymine, thymine-adenine, cytosine-guanine and guanine-cytosine. This message is transcribed as a mRNA molecule written in a different four-letter alphabet *viz.* adenine, uracil, cytosine and guanine. From a mathematical viewpoint, the transcription process is isomorphic and could therefore be reversible. This is in fact what has been experimentally demonstrated.

The *mRNA → protein* transition is a more complex process that cannot be considered isomorphic. The number of letters of the alphabet changes during this process. The genetic message, as expressed in the DNA and RNA molecules, is written in a four-letter alphabet whereas, in the protein, it is expressed in a twenty-letter alphabet. This situation shows that if the letters of the mRNA message were taken two-by-two one could specify, at most, $4^2 = 16$ different aminoacids. This means that three bases are required to define a functional codon. Hence it appears that a codon is a triplet of bases located on the tRNA loop. But then one should be able to specify $4^3 = 64$ different aminoacids and we know that proteins are made up of at most 20 different aminoacids. This means that the code is degenerate *viz.* that several different triplets could code for the same aminoacid, or that some triplets are nonsense. The main implication of this reasoning is that the information transfer from mRNA to protein should be irreversible. Crick [8, 9] has coined the expression *central dogma of molecular biology* to express the idea that the transfer $\Omega_{RNA} \rightarrow \Omega_{prot}$ is irreversible. This expression is misleading for two reasons. First, it is neither a dogma nor a biological property, but the property of any code in which the alphabet of the source is larger than the alphabet of the destination. Second, it is not a dogma but the property of a pair of probability spaces devoid of isomorphism.

A priori, one can imagine two different types of codes, namely overlapping and non-overlapping. In the first one, two successive codons have one or two elements in common. In the second one they do not. In fact these two possibilities can be distinguished on the basis of straightforward experiments. If the genetic code were displaying an overlap between codons, one should expect that, upon a punctual mutation of DNA, two or three aminoacids could be replaced in the polypeptide chain (Fig. **3**). This is not what is obtained experimentally, upon replacing one base pair by another one in the DNA. Under these conditions, only one aminoacid is replaced in the protein. This clearly shows that the genetic code is a non overlapping one.

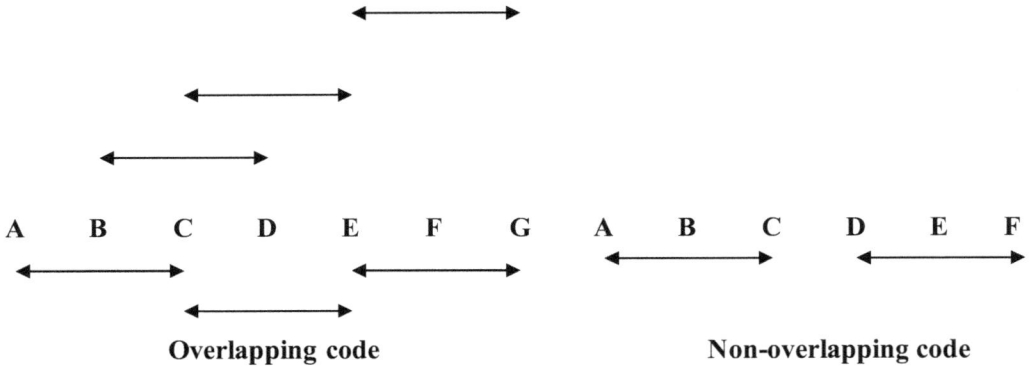

Figure 3: Overlapping and non-overlapping codes.

Some mutations, called deletions, or insertions, involve the loss, or the gain, of only one aminoacid. If the code were displaying overlaps between codons one should expect only a local perturbation of the aminoacid sequence. If, alternatively, the code displayed no overlap between codons, one should expect a scrambling of the polypeptide chain sequence (Fig. **4**). This is, in fact, what is observed.

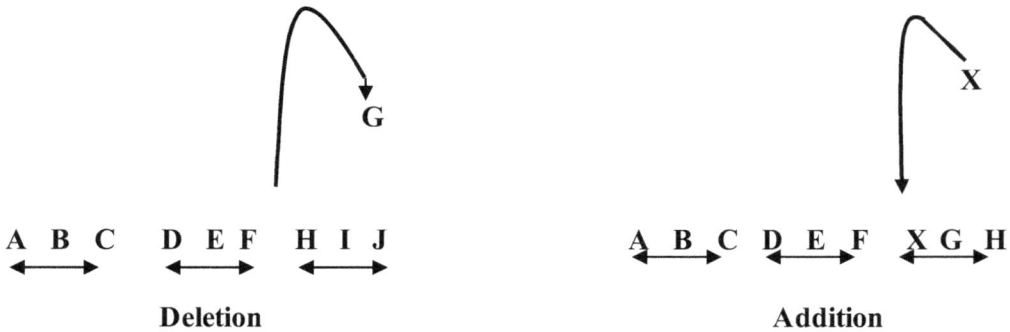

Figure 4: Effects of punctual mutations (deletion and addition) for overlapping codes.

Punctual mutations (deletions or additions) result in the scrambling of the code.

Any communication process implies that some information is lost in the channel. The information loss is called the noise. This is in fact the very basis of a principle termed the subadditivity principle. This principle requires that the sum of the self-informations $H(X)$ and $H(Y)$ of the two alphabets, X and $Y,$ be larger than, or at most equal to, the information $H(X,Y)$ which is transferred in the communication channel. This is the basis of a fundamental principle of classical communication theory called the subadditivity principle which states that

$$H(X)+H(Y) \geq H(X,Y) \tag{2.17}$$

3- GENERAL SIGNIFICANCE OF CONDITIONAL INFORMATION

Conditional information, as defined in Section 1, cannot be identified to the conditional information of the classical Shannon information theory. The conditional information $h(x_i|y_j)$, for instance, implies that y_j may physically interact with x_i, possibly changing its properties. Whereas negative $I(X:Y)$ values are prohibited in classical Shannon communication theory, negative $i(x_i:y_j)$ values are not if we assume that x_i and y_j are, for instance, molecules that can physically interact. If such a situation occurs, as postulated in Section 1 of the present Chapter, then emergence becomes possible and is antagonistic of a process of conduction of information.

If we apply the concept of information to molecules then their probabilities of occurrence on different energy levels are distributed according to the classical Gibbs-Boltzmann statistics. One has

$$p(x_i) = \exp\left\{-(E_i - E_0)/k_B T\right\} \tag{2.18}$$

where E_i and E_0 are energy levels of this molecule, k_B the Boltzmann constant and T the absolute temperature, respectively. It follows from this relationship that $h(x_i)$ has an important physical meaning that it does not possess in classical communication theory. One has

$$h(x_i) = \frac{E_i - E_0}{k_B T} \tag{2.19}$$

We can imagine the energy difference $E_i - E_0$ to be smaller for $h(x_i)$ than for $h(x_i|y_j)$. It follows then that

$$h(x_i) - h(x_i | y_j) < 0 \qquad (2.20)$$

Such a situation can be obtained, for instance, by a decrease of the E_0 value of $h(x_i)$ owing to the interaction between x_i and y_j. As we shall see, under these conditions, emergent phenomena may take place even though their occurrence is excluded from Shannon communication theory.

4- PROTEIN LATTICES COMMUNICATE OR GENERATE INFORMATION

Let us consider a protein that can bind several molecules of two types of ligands, x and y [10, 11]. The corresponding lattice is represented by a set of nodes, Ω_N, defined by the following relationship

$$\Omega_N = \left\{ p(N_{\kappa,\lambda}); \kappa, \lambda \in Z^+, \kappa, \lambda \le n \right\} \qquad (2.21)$$

In this expression, $p(N_{\kappa,\lambda})$ represents the probabilities of occurrence of the nodes of the lattice. Any node is associated with κ molecules of x and λ molecules of y. As κ and λ can take the successive values *0, 1, 2,...,n*, Ω_N collects all the nodes of the lattice. An equivalent way to describe the lattice is to replace the probabilities of occurrence of the nodes by the corresponding $h(x_{\kappa,\lambda})$ values. As a matter of fact one has

$$h(x_{\kappa,\lambda}) = -\log p(N_{\kappa,\lambda}) \qquad (2.22)$$

and the lattice can be redefined in terms of the energies of its nodes. One has then

$$\Omega_N = \left\{ h(x_{\kappa,\lambda}); \kappa, \lambda \in Z^+, \kappa, \lambda \le n \right\} \qquad (2.23)$$

Whatever the mode of description of the lattice, one can distinguish in its topology three subsets. For instance, from expression (2.21) we can define the following subsets $\Omega_0, \Omega_{Nx}, \Omega_{Ny}$. Ω_0 collects the probabilities that some protein molecules have bound neither x nor y. Moreover Ω_{Nx} and Ω_{Ny} are defined as

$$\Omega_{Nx} = \left\{ p(N_{i,\lambda}); i \in N, \lambda \in Z^+, i, \lambda \le n \right\} \qquad (2.24)$$

$$\Omega_{Ny} = \left\{ p(N_{\kappa,j}); \kappa \in Z^{+}, j \in N, \kappa, j \leq n \right\}$$

Ω_0 collects the probabilities that the protein has bound neither x nor y. We can also define the probability that a set of nodes has bound i molecules of x whether or not it has bound molecules of y. One has

$$p(x_i) = \sum_{\lambda} p(N_{i,\lambda}) \tag{2.25}$$

Similarly, the probability that a set of nodes has bound j molecules of y is

$$p(y_j) = \sum_{\kappa} p(N_{\kappa,j}) \tag{2.26}$$

The nodes bearing both x_i and y_j are collected in the probability space Ω_{XY}, namely

$$\Omega_{XY} = \left\{ p(x_i, y_j); i, j \in N \right\} \tag{2.27}$$

Such a lattice is depicted in Figs. **5** and **6**.

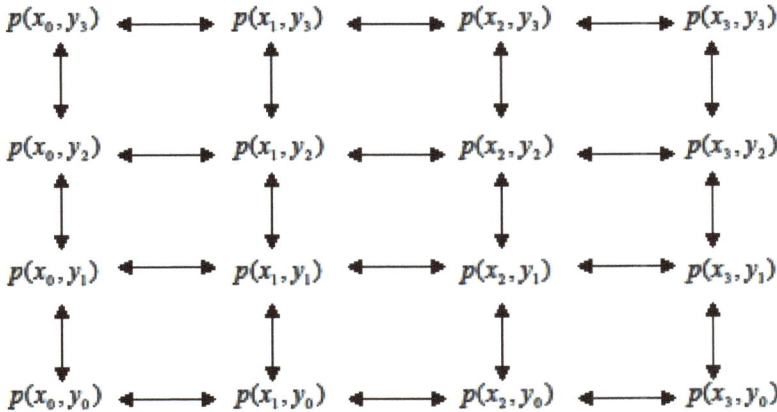

Figure 5: The probability lattice.

The protein lattice describes the successive binding of x and y molecules to the protein, or to the protein aggregate. One can define, for such a system, the usual functions involved in a communication process *viz.*

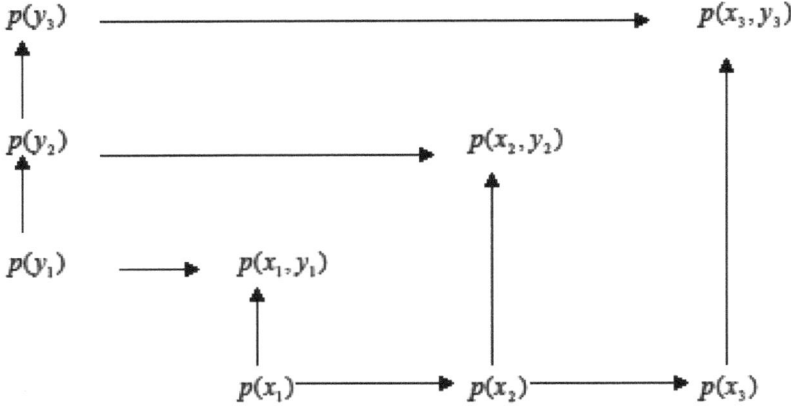

Figure 6: Connections between individual probabilities in the lattice that generate a probability channel.

$$H(X) = \sum_i \sum_j p(x_i, y_j) h(x_i) \tag{2.28}$$

$$H(Y) = \sum_i \sum_j p(x_i, y_j) h(y_j)$$

$$H(X|Y) = \sum_i \sum_j p(x_i, y_j) h(x_i | y_j)$$

$$H(Y|X) = \sum_i \sum_j p(x_i, y_j) h(y_j | x_i)$$

Hence one can raise the question to know whether such a lattice can be considered a communication channel. In practice, the system will behave as a channel if the joint probabilities such as $p(x_1, y_1), p(x_2, y_2), p(x_3, y_3)$ (Fig. **6**) and their corresponding informations are dominant in the lattice (Fig. **7**).

The important question which is at stake is to know what is the *general* condition required for the lattice to play the part of a communication channel. As already mentioned, the mathematical rule which is the very basis of any communication process is known as the subadditivity principle [3, 7]. This principle can be viewed as some kind of a switch between two different functions, namely the communication of a message and the emergence of information. Equation (2.13) above can be rewritten as

$$I(X:Y) = \sum_i \sum_j p(x_i, y_j) \log \left\{ \frac{p(x_i \mid y_j)}{p(x_i)} \right\}$$

(2.29)

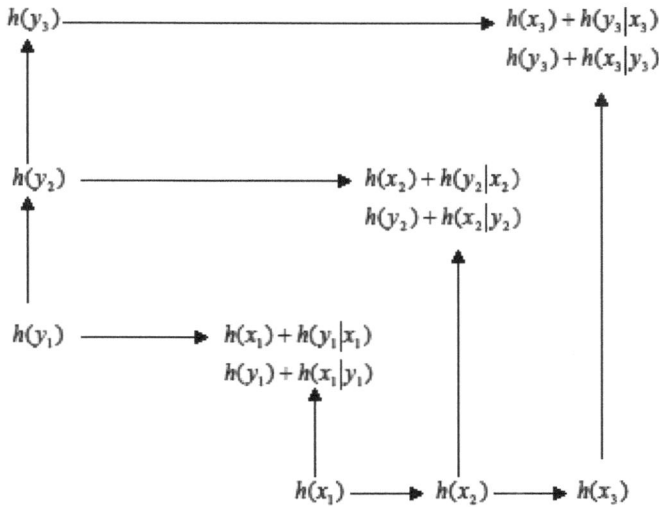

Figure 7: Connections between individual informations in the lattice that generate an information channel.

In order to find out the conditions that give rise to a communication, or to an emergence process, one has to define another function, $I^*(X:Y)$, that possesses values smaller than, or at most, equal to $I(X:Y)$. Moreover the expression of $I^*(X:Y)$ is chosen as to be equal to zero if subadditivity applies. In other words, subadditivity is the condition that results in the expression $I^*(X:Y) = 0$. In order to meet this requirement, one can notice that for all $x > 0$,

one has

$$x - 1 \geq \ln x$$

(2.30)

This expression can be rewritten as

$$x - 1 \geq \frac{1}{M} \log x$$

(2.31)

with $M = \log e$. It then follows that

$$\log e(x-1) \geq \log x \tag{2.32}$$

Hence if

$$x = \frac{p(x_i)}{p(x_i|y_j)} \tag{2.33}$$

expression (2.32) becomes

$$\log e\left\{\frac{p(x_i)}{p(x_i|y_j)} - 1\right\} \geq \log \frac{p(x_i)}{p(x_i|y_j)} \tag{2.34}$$

Hence the function $I^*(X:Y)$ can be written as

$$I^*(X:Y) = -\log e \sum_i \sum_j p(x_i,y_j)\left\{\frac{p(x_i)}{p(x_i|y_j)} - 1\right\} \tag{2.35}$$

and can be compared to the expression of $I(X:Y)$, namely

$$I(X:Y) = -\sum_i \sum_j p(x_i,y_j)\log\frac{p(x_i)}{p(x_i|y_j)} \tag{2.36}$$

It appears that

$$I(X:Y) \geq I^*(X:Y) \tag{2.37}$$

Moreover one can rewrite expression (2.35) as

$$I^*(X:Y) = -\log e \sum_i \sum_j \left\{\frac{p(x_i,y_j)p(x_i)}{p(x_i|y_j)} - p(x_i,y_j)\right\} \tag{2.38}$$

Taking account of the fact that the conditional probability $p(x_i|y_j)$ can be rewritten as

$$p(x_i|y_j) = \frac{p(x_i, y_j)}{p(y_j)} \tag{2.39}$$

under these conditions expression of $I^*(X:Y)$ becomes

$$I^*(X:Y) = -\log e \sum_i \sum_j \{p(x_i)p(y_j) - p(x_i, y_j)\} \tag{2.40}$$

or

$$I^*(X:Y) = \log e \sum_i \sum_j \{p(x_i, y_j) - p(x_i)p(y_j)\} \tag{2.41}$$

Hence the condition that generates $I^*(X:Y) = 0$, *viz.* the subadditivity condition, is

$$\sum_i \sum_j p(x_i, y_j) - \sum_i p(x_i) \sum_j p(y_j) = 0 \tag{2.42}$$

Multiplying the left-hand side member by $\log p(x_i, y_j)$ and the right-hand side member by $\log p(x_i) + \log p(y_j)$ yields

$$H(X) + H(Y) = H(X,Y) \tag{2.43}$$

which implies that the system can be considered a "perfect" communication channel.

We are now in the position to define both the conditions required for the system to behave as a communication channel or, alternatively, to generate information. It has already been shown that the condition required to behave as a communication channel is

$$\sum_i \sum_j p(x_i, y_j) - \sum_i p(x_i) \sum_j p(y_j) \geq 0 \tag{2.44}$$

This expression can be rewritten as

$$\sum_i \sum_j p(x_i)p(y_j|x_i) - \sum_i p(x_i) \sum_j p(y_j) \geq 0 \tag{2.45}$$

or as

$$\sum_i p(x_i) \left\{ \sum_j p(y_j | x_i) - \sum_j p(y_j) \right\} \geq 0 \tag{2.46}$$

The network will behave as a communication channel if

$$\sum_j p(y_j | x_i) \geq \sum_j p(y_j) \tag{2.47}$$

which is equivalent to

$$\sum_j h(y_j) \geq \sum_j h(y_j | x_i) \tag{2.48}$$

Alternatively, the network will generate information if

$$\sum_j p(y_j) - \sum_j p(y_j | x_i) > 0 \tag{2.49}$$

equivalent to

$$\sum_j h(y_j | x_i) > \sum_j h(y_j) \tag{2.50}$$

This implies that the interactions between x_i and y_j generate *novel* properties and the system becomes *emergent*. In other words *emergence* of information has replaced simple *conduction* of information. One can notice that expression (2.50) is equivalent to

$$H(X | Y) > H(X) \tag{2.51}$$

and this implies in turn that

$$I(X : Y) < 0 \tag{2.52}$$

It is evident that relations (2.51) and (2.52) cannot exist in Shannon communication theory. If it were so this would imply that the communication channel generates its own information. There is no doubt, however, that a simple biochemical lattice can possess the ability to generate information.

5- INFORMATION AND SPATIAL ORGANIZATION OF A SYSTEM

It appears from the previous results that the information of a system $i(x_A : x_B : x_C : ....)$ can be expressed as

$$i(x_A : x_B : x_C : ...) = \Sigma_i h(x_i) - \Sigma_i h(x_i | x_k, x_l, ...) \qquad (2.53)$$

A negative value of $i(x_A : x_B : x_C : ...)$ means that part of the information is used to generate spatial organization of the system. Let us consider, for example, a system of a given probability of occurrence $p(A, B, C, D, E, F)$. If, for instance, the system is organized in such a way that the probability of occurrence of A is dependant upon both B and F, the probability of occurrence of B is dependant upon A and C etc...one has

$$p(A, B, C, D, E, F) =$$
$$p(A|B, F) p(B|AC) p(C|B, D) p(D|C, E) p(E|D, F) p(F|E, A) \qquad (2.54)$$

One can write from expression (2.53)

$$\sum_i h(x_i) = h(x_A) + h(x_B) + h(x_C) + h(x_D) + h(x_E) + h(x_F)$$

and

$$h(x_A, x_B, x_C, x_D, x_E, x_F) = h(x_A | x_B, x_F) + h(x_B | x_A, x_C) + h(x_C | x_B, x_D) + h(x_D | x_C, x_E) \qquad (2.55)$$
$$+ h(x_E | x_D, x_F) + h(x_F | x_E, x_A)$$

It follows that

$$i(x_A : x_B : x_C : x_D : x_E : x_F) = h(x_A) - h(x_A | x_B, x_F) + h(x_B) - h(x_B | x_A, x_C)$$

$$+ h(x_C) - h(x_C | x_B, x_D) + h(x_D) - h(x_D | x_C, x_E) + h(x_E) - h(x_E | x_D, x_F) \qquad (2.56)$$

$$+h(x_F) - h(x_F | x_E, x_A)$$

The expression of $-i(x_A : x_B : x_C : x_D : x_E : x_F)$ represents the amount of information used to generate the organization of the system. One can notice that the system is a cyclic one (Fig. **8**).

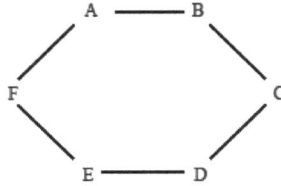

Figure 8: The topological structure of a system defined by its probability of occurrence.

The probability of occurrence of the system is (see text).
$$p(A,B,C,D,E,F) = p(A|B,F)p(B|A,C)p(C|B,D)p(D|C,E)p(E|D,F)p(F|E,A)$$

The system is shown in the Figure.

6- GENERAL CONCLUSIONS

Classical Shannon communication theory, as applied to biological systems, implies there is usually a loss of information in a communication channel but never production of information. In terms of the molecular approach of biology, this means that all the information required for the building up of a living organism is already present in the DNA molecule. This statement can be viewed as one of the central concepts of molecular biology. In this perspective, there cannot exist any emergence of information in biological systems. Such a statement is probably too simplistic to match biological reality. Simple biochemical systems, or biological lattices, generated by ligand binding to protein aggregates, or to multisited proteins may generate, or conduct, information. A protein lattice, for instance, can be considered integrated, or emergent, depending on the respective values of $\sum h(x_i)$ and $\sum h(x_i | y_j)$. Simple biochemical lattices generated by ligand binding to protein aggregates can behave either as communication channels or as generators of information. In Fig. **9** is shown a geometric representation of the functions h in the case of conduction and in the case of emergence of information.

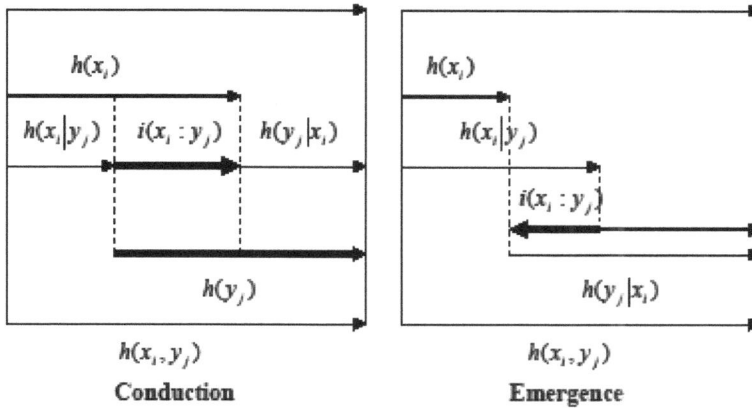

Figure 9: Positive (left) and negative (right) $i(x_i : y_j)$ values, that lead to the conduction, or emergence, of information in a system.

It is clear from the geometric representation of Fig. **8**, and its mathematical description, that the information $i(x_i : y_j)$ can be positive or negative in such a way that the whole system does not behave as a simple communication channel but, in certain circumstances, as a generator of information. We shall discuss this matter in the forthcoming Chapters.

REFERENCES

[1]   Shannon, C. E. (1948) A mathematical theory of communication. Bell System Technical Journal, 27, 379-423.

[2]   Shannon, C. E. (1948) A mathematical theory of communication. Bell System Technical Journal, 27, 623-656.

[3]   Shannon, C. E. (1949) The Mathematical Theory of Communication. University of Illinois Press, Urbana.

[4]   Kullback, S. (1959) Information Theory and Statistics.Wiley and Sons, New York.

[5]   Fano, R.M. (1961) The Transmission of Information. MIT Press and Wiley and Sons, New York.

[6]   Cover, T. M. and Thomas, J. A. (1991) Elements of Information Theory. Wiley and Sons, New York.

[7]   Callager, R.G. (1964) Information Theory. In Margenau H. and Murphy, H. eds.The Mathematics of Physics and Chemistry, Vol. II, 190-248.

[8]   Crick, F.H.C., Barnett, L., Brenner, S. and Watts-Tobin, R. J. (1961) General nature of the genetic code for proteins. Nature 192, 1227-1232.

[9]   Crick, F.H.C. (1966) The genetic code III. Scientific American 215(4), 55-62.

[10]   Ricard, J. (2006) Emergent Collective Properties, Networks and Information in Biology. Elsevier, Amsterdam, Boston.

[11]   Ricard, J. (2011) Biochemical lattices and networks as models of living systems: a problem of artificial life. Trends in Cell and Molecular Biology, 6, 9-30.

Send Orders for Reprints to reprints@benthamscience.net
Biological Systems: Complexity and Artificial Life, 2014, 35-58   35

CHAPTER 3

Emergence of Information in Biological Systems

Abstract: The concept of *emergence* is absent from classical molecular biology for it is implicitly, or explicitly, assumed that all the properties of living systems are *in fine* carried out by specific macromolecules. The concept of *biological system* implies that a number of biological properties are not borne by specific macromolecules but by a system of macromolecules. In this perspective a biological property originates from the *connections* between molecules belonging to a system. In classical molecular biology a property does not *emerge* for it is carried by a specific macromolecule. In systems biology properties *emerge* from the interactions that take place between the elements of a system.

Keywords: Information of biological systems, Joint probability, Boltzmann statistics, Allostery, Induced fit, Information and induced fit, Information and ambiguity, Information and departure from thermodynamic equilibrium, Functions of connection, Generalized microscopic reversibility.

The concept of emergence does not seem to belong to the classical domain of life sciences that seems to remain, to a large extent, analytic and reductionist in their essence. Classical molecular biology, for instance, attempts at studying biological phenomena at the macromolecular level and neglects the possible existence of emergence of novel properties that might be generated through the interaction between different elements of the system under study. In this perspective, biological functions should be specifically borne by some molecular entities and should not result from the interactions between them.

Within the frame of systems biology, it appears more and more evident that important biological functions are not specifically borne by molecular structures but, "emerge" from the transient interactions that exist between molecular entities. We have seen, in the previous Chapter, that emergence of information, often accompanied by the emergence of a function, exists. The aim of the present Chapter is to discuss some functional implications of the concept of emergence.

1- THE CONCEPT OF EMERGENCE OF INFORMATION IN SIMPLE BIOLOGICAL SYSTEMS

In Chapter 2, we have proposed a definition of the concept of information as the property that gives a material entity its identity *i.e.* the ability to be discriminated among similar entities. This definition can be made more general and can be applied to networks. In Fig. **1**, are shown some networks that can be considered simple models of possible biological processes. In Fig. **1A** is shown a linear network in which the nodes do not display mutual retroaction effects. In the second network (**1B**) it is assumed that the nodes display mutual retroaction whereas in the third (**1C**) model diverse possible types of interactions are shown.

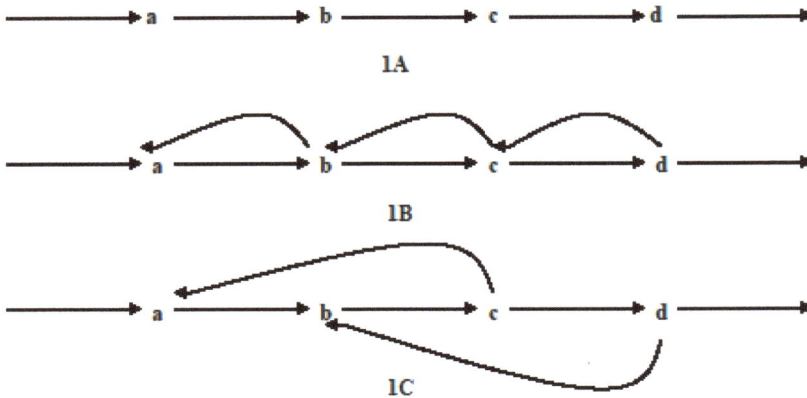

Figure 1: The probability of occurrence of a linear system contributes to define its organization.

One can derive the joint probability of occurrence, $p(a,b,c,d)$, of any of these simple models. As there is no feed-back for model **1A** one has

$$p(a,b,c,d) = p(a)p(b)p(c)p(d) \tag{3.1}$$

In the case of model **1B**, one can notice that b interacts with a, but that c interacts with b, and d interacts with c. It follows that

$$p(a|b,c,d) = p(a|b) \tag{3.2}$$

Similarly c interacts with b but not with d. Hence one has

$$p(b|c,d) = p(b|c) \tag{3.3}$$

Hence the probability of occurrence of system **1B** is

$$p(a,b,c,d) = p(a|b)p(b|c)p(c|d)p(d) \tag{3.4}$$

The situation is different for the next model. For model **1C**, *b* and *d* do not interact with *a*, hence

$$p(a|b,c,d) = p(a|c) \qquad (3.5)$$

Similarly, for the same model, *c* does not interact with *b*, therefore

$$p(b|c,d) = p(b|d) \qquad (3.6)$$

The probability of occurrence of system **1C** is then

$$p(a,b,c,d) = p(a|c)p(b|d)p(c)p(d) \qquad (3.7)$$

From the results presented in the previous Chapter it appears that the information spontaneously generated by a network can be expressed as

$$i(x_1 : x_2 : ... : x_k) = \sum_i h(x_i) - h(x_1, x_2, ..., x_k) \qquad (3.8)$$

$h(x_1), h(x_2), ...$are the informations already present in the network upon assuming that x_1, x_2 .do not interact. $\sum_i h(x_i)$ is derived from the joint probability

$$p(x_1, x_2, ...) = p(x_1)p(x_2).... \qquad (3.9)$$

and from the relationship

$$h(x_i) = -\log p(x_i) \qquad (3.10)$$

The term $h(x_1, x_2, ..., x_k)$ in expression (3.8) describes how the elements x_i and x_{i+j} interact. Thus, for instance, if x_2 interact with x_1, x_3 with x_2 etc…one will have

$$h(x_1, x_2, ..., x_k) = \sum h(x_i|x_{i+1}) \qquad (3.11)$$

Let us consider the two pathways shown in (Fig. **2A**) and (Fig. **2B**) both of them display feed-back processes. In the process **2A**, *e* interacts with the conversion process of *a* to *b*. In the process **2B**, *c* modifies the rate of conversion of *a* to *b* and *e* modifies the rate of conversion of *c* to *d*.

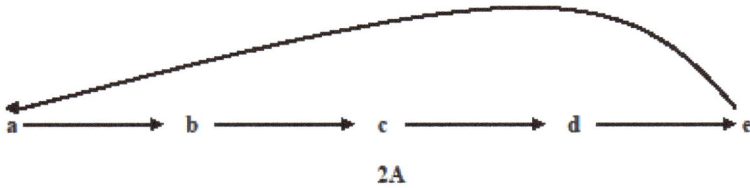

2A

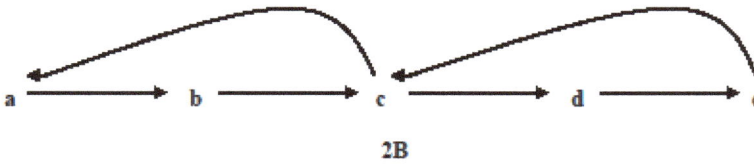

2B

Figure 2: Feedback processes contribute to the definition of the organization of a system.

If we follow the reasoning developed above, it appears that the first term $\sum_i h(x_i)$ of equation (3.8) is

$$\sum_i h(x_i) = h(a) + h(b) + h(c) + h(d) + h(e) \tag{3.12}$$

and the second term of the same equation can be written as

$$h(x_1, x_2, ...) = h(a|e) + h(b) + h(c) + h(d) + h(e) \tag{3.13}$$

and equation (3.8) above becomes

$$i(x_1 : x_2 : ....) = \sum_i h(x_i) - h(x_1, x_2, ...) = h(a) - h(a|e) \tag{3.14}$$

The same reasoning applies to the second model of Fig. **2**. One has

$$\sum_i h(x_i) = h(a) + h(b) + h(c) + h(d) + h(e) \tag{3.15}$$

and

$$h(x_1, x_2, ...) = h(a|c) + h(b) + h(c|e) + h(d) + h(e) \tag{3.16}$$

It follows that

$$i(x_1 : x_2 : ...) = \sum_i h(x_i) - h(x_1, x_2, ...) = h(a) - h(a|c) + h(c) - h(c|e) \tag{3.17}$$

If, in equations (3.14) and (3.17), $h(a|e) > h(e)$, $h(a|c) > h(a)$ and $h(c|e) > h(c)$ then the functions $i(x_1 : x_2 : ....)$ (equations 3.14 and 3.17) adopt negative values and, for both systems of Fig. **2**, feedback inhibition generates information.

This reasoning can be extended to many different biochemical networks. The analysis can be performed on the basis of the equation

$$i(a:b:c:d) = h(a) + h(b) + h(c) + h(d) - h(a,b,c,d) \tag{3.18}$$

For systems 1.A to 1.E one finds

$$i(a:b:c:d) = 0 \tag{1.A}$$

$$i(a:b:c:d) = h(a) - h(a|b) + h(b) - h(b|c) + h(c) - h(c|d) \tag{1.B}$$

$$i(a:b:c:d) = h(a) - h(a|c) + h(b) - h(b|d) \text{ (1.C)} \tag{3.19}$$

$$i(a:b:c:d) = h(a) - h(a|b,c,d) \tag{1.D}$$

$$i(a:b:c:d) = h(a) - h(a|d) + h(b) - h(b|d) + h(c) - h(c|d) \tag{1.E}$$

One can observe that the systems $(1.A) - (1.E)$ can spontaneously generate information if the conditional informations such as $h(a|b,c,d)$ or $h(b|d)$ can reach higher values than the corresponding "regular" informations, $h(a)$ and $h(b)$. In order to answer this question it is essential to come back to the physical significance of information.

2– THE PHYSICAL SIGNIFICANCE OF INFORMATION AND EMERGENCE

We have previously outlined that the probability of occurrence of molecules over different energy levels can be described by the Boltzmann statistics. One has then

$$p(x_i) = \exp\left\{-\frac{E_0 - E_i}{k_B T}\right\} \tag{3.20}$$

where E_0 and E_i are the zero and the i-th energy levels of these molecules, k_B and T the Boltzmann constant and the absolute temperature, respectively. From this definition, it follows that

$$h(x_i) = -\frac{E_0 - E_i}{k_B T} \tag{3.21}$$

In fact, the information $h(x_i)$ is the energy required by the system to jump from the E_0 level to the E_i level. This situation can be extended to a set of interacting molecules, for instance x_i and y_j. In this case, the corresponding information, $h(x_i|y_j)$, can be expressed in terms of an energy difference, $-(E_{0j} - E_{ij})$. One has then

$$h(x_i|y_j) = -\frac{E_{0j} - E_{ij}}{k_B T} \tag{3.22}$$

If the interaction between x_i and y_j is such that

$$E_{ij} - E_{0j} > E_i - E_0 \tag{3.23}$$

it follows that the interaction between x_i and y_j generates emergence of information *i.e.* emergence of energy. This situation is depicted in Fig. **3**.

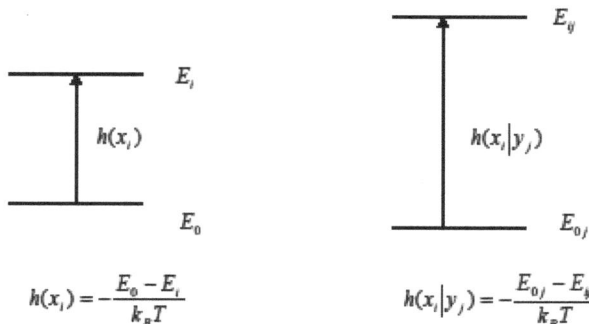

$$h(x_i) = -\frac{E_0 - E_i}{k_B T} \qquad h(x_i|y_j) = -\frac{E_{0j} - E_{ij}}{k_B T}$$

Figure 3: A process of emergence.

The interaction between x_i and y_j can generate an increase of the energy level of x_i.

This reasoning can be easily extended to more complex situations such as the one described in Fig. **1D** where different substances b, c and d can alter the behavior of a. If $h(a|b,c,d) > h(a)$, the corresponding system displays an emergent behaviour (Fig. **4**). It then appears that information can spontaneously emerge in biochemical networks.

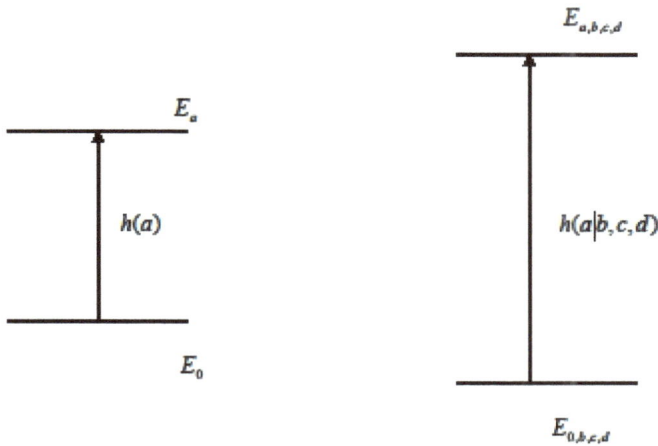

Figure 4: Several ligands (b,c,d) can contribute to a process of emergence.

The interaction of b, c, d can increase the amplitude of the energy transistion.

3- INFORMATION, ALLOSTERY AND INDUCED-FIT IN PROTEIN SYSTEMS

Proteins, for instance enzymes, are able to bind small molecules. In the case of enzymes, this binding process is accompanied by a conformation change of the protein required for its catalytic activity. A priori, one could postulate that this conformation change occurs spontaneously, or that it is induced by ligand binding to the protein. In the first case the enzyme is said to follow an allosteric mechanism. In the second case, it is assumed that the collision of the enzyme with the ligand induces a conformation change of the enzyme required for its activity. These two assumptions have important implications. In the first case the activity on an enzyme is, in a way, written in its structure. In the second case, enzyme activity is generated by the collision of the protein with its substrate. Induction of such an activity is in fact the *emergence* of a biological function, or of an *information*, owing to the collision of the protein with the ligand.

3.1- Allostery

Monod, Wyman and Changeux [1, 2] have proposed the view that certain proteins, called allosteric, exist, in the absence of any ligand, under different conformations in equilibrium. A ligand, a substrate in the case of an enzyme, could bind preferentially to one of these conformation states and therefore would shift the enzyme pre-equilibrium towards one of these states. The result of such a situation is that the substrate-binding curve is sigmoidal and characteristic of a positive co-operativity. It then appears that ligand binding is not accompanied with the emergence of any novel property for the free protein is assumed to exist under two conformations in equilibrium. In this perspective one cannot conclude that ligand binding to the protein has generated the emergence of a new protein conformation for this conformation was already present, possibly in minute amounts, in the absence of the ligand.

3.2- Induced Fit

In many cases, however, it is not this type of co-operativity which is observed but another one called negative co-operativity [3-10]. This type of co-operativity has an explanation completely different from the previous one. It implies that ligand binding to the protein *induces* its change of conformation. In such a model, the induced conformation change is a *new* process induced by the collision of the ligand and the protein. This model is more general than the previous one for it may generate positive, or negative, co-operativity as well as mixed positive-negative co-operativity. The aim of the present Section of this Chapter is to show that protein co-operativity can be identified to a process of *emergence* as defined in the previous Sections of this Chapter.

Let us consider a process of ligand binding to a tetrameric protein. Let assume for simplicity that the process is irreversible (this is an oversimplification). One has

$$E_0 \longrightarrow E_1 \longrightarrow E_2 \longrightarrow E_3 \longrightarrow E_4$$

In this simplified scheme, E_1 means that the protein has bound one ligand molecule, E_2 means that the protein has bound two ligand molecules *etc*.... Affinity of the ligand for the protein depends upon the three-dimensional structure of this protein which is itself dependant upon the number of the ligand molecules bound to the protein. This means that the properties of a given state, say E_2 for instance, are dependant on this state but also upon the protein states that have preceded E_2, namely E_0 and E_1. One could apply to this process the reasoning followed previously and define the two functions $\sum h(x_i)$ and $h(x_1, x_2, ...)$.

One has

$$\sum_i h(x_i) = h(E_1) + h(E_2) + h(E_3) + h(E_4) \tag{3.24}$$

and

$$h(x_1, x_2, ...) = h(E_1 | E_0) + h(E_2 | E_1, E_0) + h(E_3 | E_2, E_1, E_0) + h(E_4 | E_3, E_2, E_1, E_0) \tag{3.25}$$

It is then obvious that if

$$i(x_1 : x_2 : ....) = \sum_i h(x_i) - h(x_1, x_2, ...) \tag{3.26}$$

is positive the system does not display any emergence. Alternatively, if this difference is negative it displays emergent properties. In expression (3.25), however, terms such as $h(E_2 | E_1, E_0)$ do not possess clear physical meaning. Hence it is therefore essential to find out this meaning. Let us consider a step of ligand binding to the protein

$$E_{i-1} + S \rightleftharpoons E_i$$

The energy level of the protein which is going to bind ligand S is noted $i-1$. Its energy level after it has bound S is noted i. The binding constant of S to E_{i-1} is denoted K_i. From an energy viewpoint this binding constant K_i is dependant upon two factors: the intrinsic binding constant K^* of the ligand to an ideally isolated subunit and the energy involved in the interactions between the subunits.

This second factor is strictly related to the difference between the free energies E_i and E_{i-1}. One has then

$$K_i = K^* \exp\left\{-\frac{E_{i-1} - E_i}{k_B T}\right\} \tag{3.27}$$

The ligand binding isotherm of a tetrameric protein that binds, for instance, four ligand molecules can be written as

$$v = \frac{K_1[S] + 2K_1K_2[S]^2 + 3K_1K_2K_3[S]^3 + 4K_1K_2K_3K_4[S]^4}{1 + K_1[S] + K_1K_2[S]^2 + K_1K_2K_3[S]^3 + K_1K_2K_3K_4[S]^4} \tag{3.28}$$

If we consider the products of equilibrium binding constants that appear in the numerator and denominator of the binding equation (3.28) the products $K_1K_2....K_i$ can be reexpressed as

$$K_1 = K^* \exp\left\{-\frac{E_0 - E_1}{k_B T}\right\}$$

$$K_1K_2 = K^{*2} \exp\left\{-\frac{E_0 - E_2}{k_B T}\right\} \tag{3.29}$$

$$K_1K_2K_3 = K^{*3} \exp\left\{-\frac{E_0 - E_3}{k_B T}\right\}$$

$$K_1K_2K_3K_4 = K^{*4} \exp\left\{-\frac{E_0 - E_4}{k_B T}\right\}$$

One can notice that the free energy differences in the exponential terms are, in fact, equivalent to conditional informations $h(x_i|y_j)$. One has

$$\frac{E_1 - E_0}{k_B T} = h(x_1)$$

$$\frac{E_2 - E_0}{k_B T} = h(x_1) + h(x_2|x_1) \tag{3.30}$$

$$\frac{E_3 - E_0}{k_B T} = h(x_1) + h(x_2|x_1) + h(x_3|x_1, x_2)$$

$$\frac{E_4 - E_0}{k_B T} = h(x_1) + h(x_2|x_1) + h(x_3|x_1, x_2) + h(x_4|x_1, x_2, x_3)$$

It follows from these relations that

$$\frac{E_1 - E_0}{k_B T} = K_1 = h(x_1)$$

$$\frac{E_2 - E_1}{k_B T} = K_2 = h(x_2|x_1) \tag{3.31}$$

$$\frac{E_3 - E_2}{k_B T} = K_3 = h(x_3|x_1, x_2)$$

$$\frac{E_4 - E_3}{k_B T} = K_4 = h(x_4|x_1, x_2, x_3)$$

This means that the binding constants can be expressed in terms of informations. It also appears that substrate binding to the protein can be associated with emergent effects. This is shown in Fig. **5**. One can see, in this Figure, that substrate binding to the first subunit produces an increase of energy equal to $E_1 - E_0$. Moreover the binding of the same substrate to the second subunit is associated with a larger free energy equal to $E_2 - E_1$. From relations (3.31) above this implies that

$$h(x_1) < h(x_2|x_1) \tag{3.32}$$

Moreover when the substrate binds to the third subunit, the corresponding increase of energy $E_3 - E_2$ is again larger than the preceding process. This situation implies that

$$h(x_2|x_1) < h(x_3|x_1, x_2) \tag{3.33}$$

Relations (3.32) and (3.33) show that during the binding of the substrate to the second and third subunits there is *emergence* of free energy. More generally speaking, it is remarkable that an equation of ligand binding to a protein can be expressed in terms of information and conditional information.

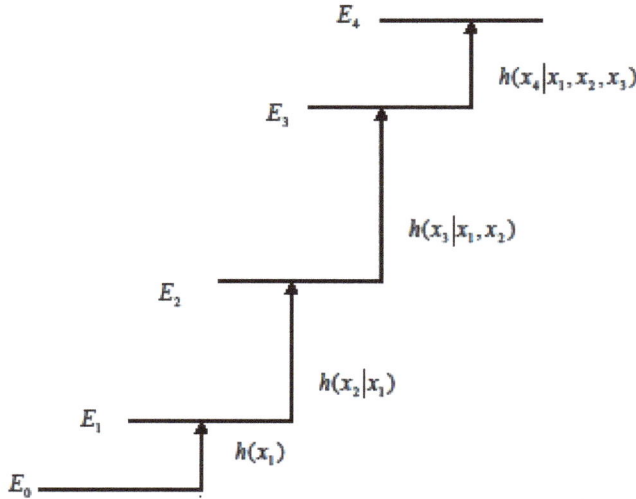

Figure 5: Emergent processes during ligand binding to a protein. Emergence of binding energy takes place, in this process, during the second and third step of ligand binding to the protein for then one has $h(x_2|x_1) > h(x_1)$ and $h(x_3|x_2,x_1) > h(x_2|x_1)$.

4- EMERGENCE OF INFORMATION AND AMBIGUITY

Emergence of information can be associated with some ambiguity. Let us consider for instance, in an enzyme or a chemical process, the same event b,c. Its probability of occurrence, $p(b,c)$, can be expressed in two different, but equivalent, ways. One has

$$p(b,c) = p(b|c)p(c) \tag{3.34}$$

$$p(b,c) = p(c|b)p(b)$$

which implies that

$$-\log p(b|c) - \log p(c) = -\log p(c|b) - \log(b) \tag{3.35}$$

It follows that

$$h(c) - h(c|b) = h(b) - h(b|c) \qquad (3.36)$$

Emergence of information is obtained if

$$h(c|b) > h(c) \qquad (3.37)$$

which implies that

$$h(b|c) > h(b) \qquad (3.38)$$

This situation is depicted in the Fig. **6**.

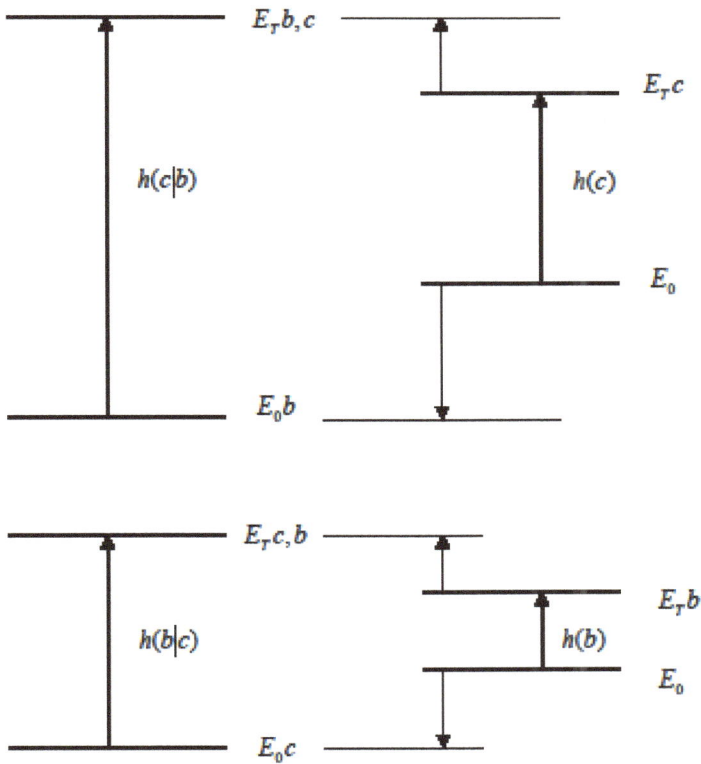

Figure 6: Reciprocity of the action of the elements, b and c, of a binary complex.

b stimulates the action of *c* and *c* stimulates the action of *b*.

The ambiguity is even more evident in the case of complex systems. Thus, for instance, the joint probability $p(a,b,c,d)$ of a chemical, or of an enzyme, reaction can be expressed in many different, but equivalent ways, for instance

$$p(a,b,c,d) = p(a|b,c,d)p(b,c,d) \qquad (3.39)$$

$$p(a,b,c,d) = p(b|a,c,d)p(a,c,d)$$

Moreover one has

$$p(b,c,d) = p(b|c,d)p(c,d) \qquad (3.40)$$

$$p(a,c,d) = p(a|c,d)p(c,d)$$

and it follows from these relationships that

$$p(a|b,c,d)p(b|c,d) = p(b|a,c,d)p(a|c,d) \qquad (3.41)$$

which can be rewritten as

$$\log p(a|b,c,d) - \log p(a|c,d) = \log p(b|a,c,d) - \log p(b|c,d) \qquad (3.42)$$

or as

$$h(a|c,d) - h(a|b,c,d) = h(b|c,d) - h(b|a,c,d) \qquad (3.43)$$

Emergence will be obtained if

$$h(a|b,c,d) > h(a|c,d) \qquad (3.44)$$

$$h(b|a,c,d) > h(b|c,d)$$

This situation is depicted in Fig. **7**

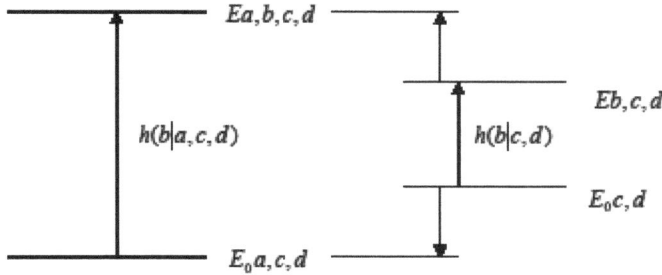

Figure 7: One of the elements (a), in a multi-molecular complex, is responsible for the enhanced activity of the complex.

5- EMERGENCE AND DEPARTURE FROM EQUILIBRIUM

5.1- Linear Networks

Let us consider a sequence of chemical reactions catalysed by a bi-enzyme complex C_0. This complex binds either substrate S_1 on enzyme E_1 or substrate S_2 on enzyme E_2. If, however, substrate S_2 has been bound first to enzyme E_2 this process prevents enzyme E_1 from binding S_1. The resulting overall process of substrate binding to the complex followed by catalysis is then linear (Fig. **8**).

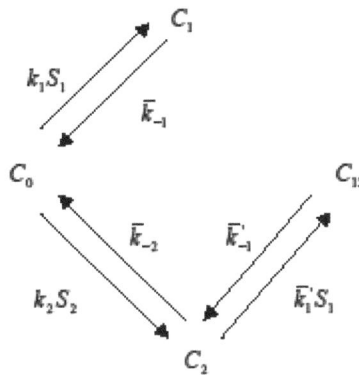

Figure 8: One route only for the binding of two ligands on a protein.

Any enzyme reaction involved in this linear network is characterized by three distinct events: substrate binding, substrate release and catalysis. These processes can be collected as an apparent affinity constant. One has

$$\overline{K}_i = \frac{k_i}{k_{-i} + k_{ci}} = \frac{k_i}{\overline{k}_{-i}} \tag{3.45}$$

where k_i is the substrate binding constant, k_{-i} the substrate release constant and k_{ci} the relevant catalytic constant. The concentrations $[C_1]$, $[C_2]$ and $[C_{12}]$ of the three enzyme complexes can be expressed as

$$\frac{[C_1]}{[C_0]} = \bar{K}_1[S_1]$$

$$\frac{[C_2]}{[C_0]} = \bar{K}_2[S_2] \qquad\qquad (3.46)$$

$$\frac{[C_{12}]}{[C_0]} = \bar{K}_1\bar{K}_2'[S_1][S_2]$$

where $[C_0]$ is the concentration of the free, unbound, complex. The existence of catalytic constants within the expressions of $\bar{K}_1, \bar{K}_2$ and $\bar{K}_2'$ prevents the system from being in equilibrium. One can derive, under these conditions, the probability of occurrence of, for instance, the S_1 state. One finds

$$p(S_1) = \frac{[C_1]/[C_0]+[C_{12}]/[C_0]}{1+[C_1]/[C_0]+[C_2]/[C_0]+[C_{12}]/[C_0]} \qquad\qquad (3.47)$$

Similarly, one can derive the conditional probability of occurrence of S and one has

$$p(S_1|S_2) = \frac{[C_{12}]/[C_0]}{[C_2]/[C_0]+[C_{12}]/[C_0]} \qquad\qquad (3.48)$$

Taking advantage of expressions (3.46) one has

$$p(S_1) = \frac{\bar{K}_1[S_1]+\bar{K}_1\bar{K}_2'[S_1][S_2]}{1+\bar{K}_1[S_1]+\bar{K}_2[S_2]+\bar{K}_1\bar{K}_2'[S_1][S_2]} \qquad\qquad (3.49)$$

and

$$p(S_1|S_2) = \frac{\bar{K}_1\bar{K}_2'[S_1]}{\bar{K}_2+\bar{K}_1\bar{K}_2'[S_1]} \qquad\qquad (3.50)$$

The information taken up, or released, through the interaction between S_1 and S_2 is then

$$\frac{p(S_1|S_2)}{p(S_1)} = \frac{\bar{K}_2' + \bar{K}_1\bar{K}_2'[S_1] + \bar{K}_2\bar{K}_2'[S_2] + \bar{K}_1\bar{K}_2'^2[S_1][S_2]}{\bar{K}_2 + \bar{K}_1\bar{K}_2'[S_1] + \bar{K}_2\bar{K}_2'[S_2] + \bar{K}_1\bar{K}_2'^2[S_1][S_2]} \tag{3.51}$$

It appears from this equation that information can be generated if $\bar{K}_2 > \bar{K}_2'$ or taken up if $\bar{K}_2 < \bar{K}_2'$.

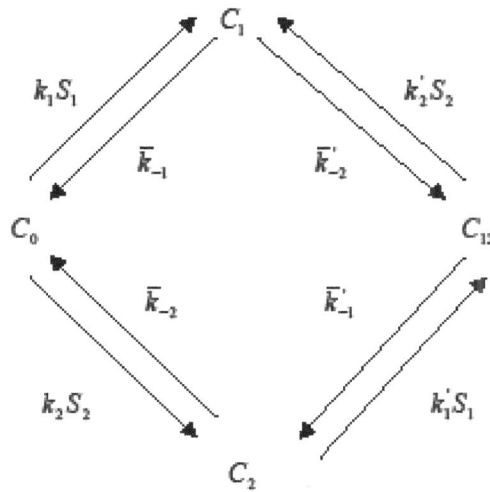

Figure 9: Two different routes for the binding of two ligands on a protein.

Now if we assume, as shown in Fig. **9**, that one can go from node C_0 to node C_{12} through two different routes then the relative concentrations $[C_1]/[C]_T$; $[C_2]/[C]_T$ and $[C_{12}]/[C]_T$ can be expressed as

$$\frac{[C_1]}{[C_0]} = \bar{K}_1[S_1] + u_1$$

$$\frac{[C_2]}{[C_0]} = \bar{K}_2[S_2] + u_2 \tag{3.52}$$

$$\frac{[C_{12}]}{[C_0]} = \bar{K}_1\bar{K}_2'[S_1][S_2] + u_{12} = \bar{K}_1'\bar{K}_2[S_1][S_2] + u_{12}'$$

If $u_1 \neq u_2$ and $u_{12} \neq u'_{12}$ the concentrations $[C_1], [C_2]$ and $[C_{12}]$ should be dependant upon the two reactions pathways leading to C_{12} except if

$$\bar{K}_1 \bar{K}'_2 - \bar{K}'_1 \bar{K}_2 = 0 \qquad (3.53)$$

This condition which applies to non-equilibrium systems has been called *generalized microscopic reversibility* [11]. As we shall see later, such a condition relies upon a physical principle and should not be considered fortuitous.

5.2- Alternative Routes and Functions of Connection

The functions u can be obtained by writing down the steady state equations for $[C_1]/[C_0], [C_2]/[C_0]$ and $[C_{12}]/[C_0]$. One obtains

$$k_1[S_1] + \bar{k}'_{-2} \frac{[C_{12}]}{[C_0]} - (\bar{k}_{-1} + k'_2[S_2]) \frac{[C_1]}{[C_0]} = 0$$

$$k_2[S_2] + \bar{k}'_{-1} \frac{[C_{12}]}{[C_0]} - (\bar{k}_{-2} + k'_1[S_1]) \frac{[C_2]}{[C_0]} = 0 \qquad (3.54)$$

$$k'_2[S_2] \frac{[C_1]}{[C_0]} + k'_1[S_1] \frac{[C_2]}{[C_0]} - (\bar{k}'_{-2} + \bar{k}'_{-1}) \frac{[C_{12}]}{[C_0]} = 0$$

These equations can be presented in matrix form and one obtains

$$\begin{bmatrix} -(\bar{k}_{-1} + k'_2[S_2]) & 0 & \bar{k}'_{-2} \\ 0 & -(\bar{k}_{-2} + k'_1[S_1]) & \bar{k}'_{-1} \\ k_2[S_2] & k'_1[S_1] & -(\bar{k}'_{-1} + \bar{k}'_{-2}) \end{bmatrix} \begin{bmatrix} u_1 \\ u_2 \\ u_{12} \end{bmatrix} = \begin{bmatrix} 0 \\ \bar{k}'_{-1}[S_1][S_2](\bar{K}_1 \bar{K}'_2 - \bar{K}'_1 \bar{K}_2) \\ -\bar{k}'_{-1}[S_1][S_2](\bar{K}_1 \bar{K}'_2 - \bar{K}'_1 \bar{K}_2) \end{bmatrix} \qquad (3.55)$$

which is equivalent to

$$\begin{bmatrix} -(\bar{k}_{-1} + k'_2[S_2]) & 0 & \bar{k}'_{-2} \\ 0 & -(\bar{k}_{-2} + k'_1[S_1]) & \bar{k}'_{-1} \\ k'_2[S_2] & k'_1[S_1] & -(\bar{k}'_{-2} + \bar{k}'_{-1}) \end{bmatrix} \begin{bmatrix} u_1 \\ u_2 \\ u'_{12} \end{bmatrix} = \begin{bmatrix} -\bar{k}'_{-2}[S_1][S_2](\bar{K}_1 \bar{K}'_2 - \bar{K}'_1 \bar{K}_2) \\ 0 \\ \bar{k}'_{-2}[S_1][S_2](\bar{K}_1 \bar{K}'_2 - \bar{K}'_1 \bar{K}_2) \end{bmatrix} \qquad (3.56)$$

It follows from this reasoning that

$$u_1 = u_2 = u_{12} = u'_{12} = 0 \tag{3.57}$$

if the conditions for generalized microscopic reversibility apply. Solving systems (3.55) and (3.56) allows to derive the expression of the functions of connection $u's$. One finds

$$u_1 = -\frac{\bar{k}_{-2}\bar{k}'_{-1}\bar{k}'_{-2}[S_1][S_2](\bar{K}_1\bar{K}'_2 - \bar{K}'_1\bar{K}_2)}{\bar{k}_{-1}\bar{k}_{-2}(\bar{k}'_{-1} + \bar{k}'_{-2}) + \bar{k}_{-1}\bar{k}'_{-2}k'_1[S_1] + \bar{k}_{-2}\bar{k}'_{-1}k'_2[S_2]}$$

$$u_2 = \frac{\bar{k}_{-1}\bar{k}'_{-2}\bar{k}'_{-1}[S_1][S_2](\bar{K}_1\bar{K}'_2 - \bar{K}'_1\bar{K}_2)}{\bar{k}_{-1}\bar{k}_{-2}(\bar{k}'_{-1} + \bar{k}'_{-2}) + \bar{k}_{-1}\bar{k}'_{-2}k'_1[S_1] + \bar{k}_{-2}\bar{k}'_{-1}k'_2[S_2]} \tag{3.58}$$

$$u_{12} = -\frac{\bar{k}'_{-1}\bar{k}_{-2}[S_1][S_2](\bar{k}_{-1} + k'_2[S_2])(\bar{K}_1\bar{K}'_2 - \bar{K}'_1\bar{K}_2)}{\bar{k}_{-1}\bar{k}_{-2}(\bar{k}'_{-1} + \bar{k}'_{-2}) + \bar{k}_{-1}\bar{k}'_{-2}k'_1[S_1] + \bar{k}_{-2}\bar{k}'_{-1}k'_2[S_2]}$$

$$u'_{12} = \frac{\bar{k}_{-1}\bar{k}'_{-2}[S_1][S_2](\bar{k}_{-2} + k'_1[S_1])(\bar{K}_1\bar{K}'_2 - \bar{K}'_1\bar{K}_2)}{\bar{k}_{-1}\bar{k}_{-2}(\bar{k}'_{-1} + \bar{k}'_{-2}) + \bar{k}_{-1}\bar{k}'_{-2}k'_1[S_1] + \bar{k}_{-2}\bar{k}'_{-1}k'_2[S_2]}$$

One can notice that the two functions of connection, u_{12} and u'_{12} are nonlinear in S_1 and S_2. The probabilities $p(S_1)$ and $p(S_1|S_2)$ can now be derived under conditions where generalized microscopic reversibility does not necessarily apply. One finds

$$p(S_1) = \frac{\bar{K}_1[S_1] + \bar{K}_1\bar{K}'_2[S_1][S_2] + u_1 + u_{12}}{1 + \bar{K}_1[S_1] + \bar{K}_2[S_2] + \bar{K}_1\bar{K}'_2[S_1][S_2] + u_1 + u_2 + u_{12}} \tag{3.59}$$

$$p(S_1|S_2) = \frac{\bar{K}_1\bar{K}'_2[S_1][S_2] + u_{12}}{\bar{K}_2[S_2] + \bar{K}_1\bar{K}'_2[S_1][S_2] + u_2 + u_{12}}$$

or as

$$p(S_1) = \frac{\bar{K}_1[S_1] + \bar{K}_1\bar{K}'_2[S_1][S_2] + u_1 + u'_{12}}{1 + \bar{K}_1[S_1] + \bar{K}_2[S_2] + \bar{K}_1\bar{K}'_2[S_1][S_2] + u_1 + u_2 + u'_{12}} \tag{3.60}$$

$$p\left(S_1|S_2\right)=\frac{\overline{K}_1\overline{K}_2'\left[S_1\right]\left[S_2\right]+u_{12}'}{\overline{K}_2\left[S_2\right]+\overline{K}_1\overline{K}_2'\left[S_1\right]\left[S_2\right]+u_2+u_{12}'}$$

Let us define from expression (3.59) new functions $u_\alpha^*, u_\beta^*, u_\gamma^*$ and u_δ^* as

$$u_\alpha^*=\frac{u_1+u_{12}}{\overline{K}_1\left[S_1\right]+\overline{K}_1\overline{K}_2'\left[S_1\right]\left[S_2\right]}$$

$$u_\beta^*=\frac{u_1+u_2+u_{12}}{1+\overline{K}_1\left[S_1\right]+\overline{K}_2\left[S_2\right]+\overline{K}_1\overline{K}_2'\left[S_1\right]\left[S_2\right]} \qquad (3.61)$$

$$u_\gamma^*=\frac{u_{12}}{\overline{K}_1\overline{K}_2'\left[S_1\right]\left[S_2\right]}$$

$$u_\delta^*=\frac{u_2+u_{12}}{\overline{K}_2\left[S_2\right]+\overline{K}_1\overline{K}_2'\left[S_1\right]\left[S_2\right]}$$

Taking advantage of these u^* functions, the expressions of $p(S_1)$ and $p\left(S_1|S_2\right)$ can be rewritten as

$$p\left(S_1\right)=\frac{\overline{K}_1\left[S_1\right]+\overline{K}_1\overline{K}_2'\left[S_1\right]\left[S_2\right]}{1+\overline{K}_1\left[S_1\right]+\overline{K}_2\left[S_2\right]+\overline{K}_1\overline{K}_2'\left[S_1\right]\left[S_2\right]}\frac{1+u_\alpha^*}{1+u_\beta^*} \qquad (3.62)$$

$$p\left(S_1|S_2\right)=\frac{\overline{K}_1\overline{K}_2'\left[S_1\right]\left[S_2\right]}{\overline{K}_2\left[S_2\right]+\overline{K}_1\overline{K}_2'\left[S_1\right]\left[S_2\right]}\frac{1+u_\gamma^*}{1+u_\delta^*}$$

or as

$$p\left(S_1\right)=p\left(\overline{S}_1\right)\frac{1+u_\alpha^*}{1+u_\beta^*} \qquad (3.63)$$

$$p\left(S_1|S_2\right)=p\left(\overline{S}_1|\overline{S}_2\right)\frac{1+u_\gamma^*}{1+u_\delta^*}$$

In these equations $p\left(\overline{S}_1\right)$ and $p\left(\overline{S}_1|\overline{S}_2\right)$ represent what would be the probability and the conditional probability of occurrence of S_1 if the system were following either quasi-equilibrium, or generalized microscopic reversibility, conditions. Hence the u^* functions express how the system behaves when it departs from these conditions.

Equations (3.63) can be rewritten as

$$\log p(S_1) = \log p(\overline{S}_1) + \log\frac{1+u_\alpha^*}{1+u_\beta^*} \tag{3.64}$$

$$\log p(S_1|S_2) = \log p(\overline{S}_1|\overline{S}_2) + \log\frac{1+u_\gamma^*}{1+u_\delta^*}$$

Moreover $\log(1+u)$ can be expanded in series and one has

$$\log(1+u) = 0.434\left(u - \frac{u^2}{2} + ...\right) \tag{3.65}$$

Equations (3.64) can be rewritten as

$$h(S_1) = h(\overline{S}_1) - 0.434\left(u_\alpha^* - u_\beta^*\right) \tag{3.66}$$

$$h(S_1|S_2) = h(\overline{S}_1|\overline{S}_2) - 0.434\left(u_\gamma^* - u_\delta^*\right)$$

The information generated, or taken up, in the system is then

$$h(S_1) - h(S_1|S_2) = h(\overline{S}_1) - h(\overline{S}_1|\overline{S}_2) + 0.434\left[\left(u_\gamma^* - u_\delta^*\right) - \left(u_\alpha^* - u_\beta^*\right)\right] \tag{3.67}$$

that can be written as

$$i(S_1 : S_2) = i(\overline{S}_1 : \overline{S}_2) + 0.434\left[\left(u_\gamma^* - u_\delta^*\right) - \left(u_\alpha^* - u_\beta^*\right)\right] \tag{3.68}$$

Hence it appears that the sign of $i(S_1 : S_2)$ is, in part, dependant upon that of $(u_\gamma^* - u_\delta^*) - (u_\alpha^* - u_\beta^*)$. Non-equilibrium, or lack of generalized microscopic reversibility, may result in emergence of information. In fact, lack of equilibrium *i.e.*

$$u_\gamma^* + u_\beta^* > u_\alpha^* + u_\delta^* \tag{3.69}$$

implies that

$$\frac{u_1 + u_2 + u_{12}}{1 + \overline{K}_1[S_1] + \overline{K}_2[S_2] + \overline{K}_1\overline{K}_2'[S_1][S_2]} > \frac{u_2 + u_{12}}{\overline{K}_2[S_2] + \overline{K}_1\overline{K}_2'[S_1][S_2]} \tag{3.70}$$

Such a situation is obtained if

$$u_1 > 1 + \overline{K}_1[S_1] \tag{3.71}$$

Under these conditions, non-equilibrium of the system, or lack of generalized microscopic reversibility, may result in the emergence of information in the system.

6- GENERALIZED MICROSCOPIC REVERSIBILITY

One may raise the important question to know whether the condition for generalized microscopic reversibility [11] that results in the suppression of the u terms in the previous equations is the consequence of some physical principle, or is purely fortuitous [11]. Let us consider two rate constants, k_{-1} and k_{c1}. As already pointed out, the first one refers to substrate release and the second one to catalysis. If the two transition states for substrate release, $S^{\neq}$, and catalysis, $X^{\neq}$, stabilize the same enzyme conformation, then a relationship appears between the two corresponding rate constants. One has then

$$\Delta G_{-1}^{\neq} = \Delta G_{-S1}^{\neq*} - U_\tau + U_\gamma \tag{3.72}$$

$$\Delta G_{C1}^{\neq} = \Delta G_{C1}^{\neq*} - U_\tau + U_\gamma$$

Here, $\Delta G_{-S1}^{\neq*}$ and $\Delta G_{C1}^{\neq*}$ are what the free energies of activation for substrate release and catalysis would be in the absence of protein-protein interactions. As it may not be the case for oligomeric enzymes and multi enzyme complexes, stabilization-destabilization energies are exerted on both the ground and the transition states. The energies are represented respectively by U_γ (ground state) and U_τ (transition state). Indeed relationships (3.73) are valid only because the two transition states stabilize the same conformation of the enzyme. The same reasoning can be applied to the two other constants $k_{-1}^{'}$ and $k_{C1}^{'}$, and one has

$$\Delta G_{-1}^{\neq'} = \Delta G_{-S1}^{\neq*} - U_{\tau}^{'} + U_{\gamma}^{'}$$ (3.73)

$$\Delta G_{C1}^{\neq'} = \Delta G_{C1}^{\neq*} - U_{\tau}^{'} + U_{\gamma}^{'}$$

The difference $\Delta G_{C1}^{\neq*} - \Delta G_{-S1}^{\neq*}$ can be derived from both equations (3.73) and (3.74). If we assume that protein-protein interactions stabilize, or destabilize, to the same extent the transition states for substrate release and catalysis then $U_{\tau} = U_{\tau}^{'}$ and $U_{\gamma} = U_{\gamma}^{'}$. It follows that

$$\Delta G_{C1}^{\neq} - \Delta G_{-1}^{\neq} = \Delta G_{C1}^{\neq'} - \Delta G_{-1}^{\neq'} = \Delta G_{-S1}^{\neq*} - \Delta G_{C1}^{\neq*}$$ (3.74)

which means that

$$\frac{k_{C1}}{k_{-1}} = \frac{k_{C1}^{'}}{k_{-1}^{'}}$$ (3.75)

This relationship implies the existence of generalized microscopic reversibility. It thus appears that generalized microscopic reversibility has real physical grounds but cannot be considered a physical principle that has to be of necessity fulfilled.

7- GENERAL CONCLUSIONS

Some general conclusions can be derived from the results presented in this Chapter. Information of a molecule is related to the energy levels of this molecule. As a consequence, emergence of information in a molecule means an increase of energy of that molecule. If the collision of several molecules results in the increase of their energy, the overall system increases its information. If one considers dynamic molecular systems displaying feedback interactions, one can expect these systems to display spontaneous emergence of information. At the molecular and supramolecular levels, there exist tight relationships between feedback effects and emergence of information.

Phenomena described under the name of "ligand induced co-operativity" can also be considered examples of emergent processes. One can possibly expect that the collision of a ligand with a protein induces the *emergence* of a new protein conformation. This emergent process can be quantitatively described in terms of the emergence of information that takes place upon the binding of a ligand to a protein.

Last, emergence of information can also be generated by the lack of steady state in a complex molecular system. Whatever that may be, there is little doubt that apparently "simple" multimolecular systems can spontaneously generate the information (the energy) required for their correct functioning.

REFERENCES

[1]   Monod, J., Changeux, J. P. and Jacob, F. (1963) Allosteric proteins and cellular control systems. J. Mol. Biol. 6, 306-329.
[2]   Monod, J., Wyman, J. and Changeux, J.P. (1965) On the nature of allosteric transitions J. Mol. Biol. 12, 88-118.
[3]   Koshland, D.E. (1973) Protein shape and biological control. Sci. Amer. 229(4), 52-64.
[4]   Koshland, D.E. (1984) Control of enzyme activity and metabolic pathways. Trends Biochem. Sci. 9, 155-159, 1984.
[5]   Levitzki, A. and Kosland, D. E. (1969) Negative cooperativity in regulatory enzymes. Proc. Natl. Acad. Sci 62, 1121-1128, 1969.
[6]   Levitzki, A. and Koshland, D.E. (1972) Role an allosteric effector. Guanosine triphosphate activation in cytosine triphosphate synthetase. Biochemistry, 11, é41-246.
[7]   Levitzki, A. and Koshland, D.E.(1972b) Ligand induced dimer-tetramer transformation in cytosine triphosphate synthetase. Biochemistry, 247-253.
[8]   Koshland, D.E., Nemethy, G. and Filmer, D. (1966) Comparison of experimental binding data and theoretical models in protein containing subunits. Biochemistry 5, 365-385.
[9]   Ricard, J., Mouttet, C. and Nari, J. (1974) Subunit interactions in enzyme catalysis: kinetic models for one-substrate polymeric enzymes. Eur. J. Biochem. 41, 479-497.
[10]   Ricard, J (1985). In Organized Multienzyme Systems G. R. Welch Ed. pp. 177-240, Academic Press.
[11]   Whitehead, E. (1970) The regulation of enzyme activity and allosteric transition. Progr. Biophys. Mol. Biol. 21, 449-456.

Send Orders for Reprints to reprints@benthamscience.net

Biological Systems: Complexity and Artificial Life, 2014, 59-67   **59**

CHAPTER 4

Non-Equilibrium, Enzyme Reactions, Self-Organization and Dynamic Properties

Abstract: Even "simple" enzyme networks can possibly display emergence of properties. These properties can be generated by a drift of the system away from equilibrium conditions. The lack of equilibrium of an enzyme system certainly plays a major importance in defining its function within the living cell.

Keywords: Phenomenological description of an enzyme reaction, Probability of occurrence of an event under non-equilibrium conditions, Drift of a reaction under quasi-equilibrium and non-equilibrium reaction, Perturbation terms, Random binding of substrates to an enzyme, Sequential binding of substrates to an enzyme, Emergent enzyme reactions, Drift from equilibrium, Thermodynamic implications of a drift from quasi-equilibrium.

Most enzyme reactions taking place in the living cell involve two substrates and two products. There exist some enzyme reactions involving three substrates and products but this is rather exceptional. Quite often, complex enzyme reactions are studied under conditions of pseudo-equilibrium. Usually, these conditions do not prevail in the living cell and there is little doubt that many enzyme reactions do not proceed under standard pseudo-equilibrium conditions. Hence it is important to know whether departure of such dynamic systems from standard pseudo-equilibrium conditions can generate novel, or unexpected, properties.

1- PHENOMENOLOGICAL DESCRIPTION OF SIMPLE ENZYME REACTIONS

As already pointed out**,** most enzymes act on two substrates A and B and generate two products P and Q. One has then

$$A + B \rightarrow P + Q$$

In such a system, A and B are conventionally called substrates, P and Q products. Initially, the reaction proceeds from left to right but, after a while, equilibrium is reached between reactants and products. There is little doubt that such a situation does not often occur in the living cell. One may then wonder whether the drift of the system from standard quasi-equilibrium conditions could not generate unexpected self-organization processes. Under "initial" conditions, catalysed reactions may follow a random order of substrate binding to the catalyst *viz.*

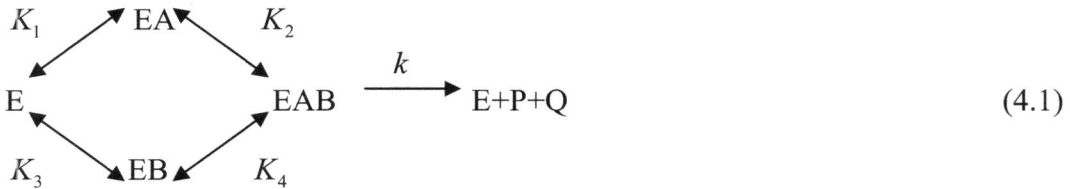

$$
\begin{array}{c}
K_1 \nearrow \text{EA} \searrow K_2 \\
\text{E} \qquad\qquad \text{EAB} \xrightarrow{\ k\ } \text{E+P+Q} \\
K_3 \searrow \text{EB} \nearrow K_4
\end{array}
\tag{4.1}
$$

However, the enzyme process may also follow a compulsory order of substrate binding. One can have for instance

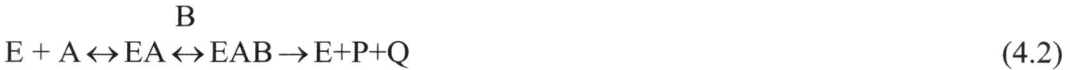

$$
\begin{array}{c}
\text{B} \\
\text{E} + \text{A} \leftrightarrow \text{EA} \leftrightarrow \text{EAB} \rightarrow \text{E+P+Q}
\end{array}
\tag{4.2}
$$

Other reaction processes could also take place that will not be considered here. For models (4.1) and (4.2), one can derive the probabilities that the enzyme has bound A, B or both. One can express these probabilities under quasi-equilibrium, or under conditions that tend to drift the system away from quasi-equilibrium. One finds for a random binding of the substrates to the enzyme under non-equilibrium conditions [1-4].

$$
p(A) = \frac{K_1[A]\big(1+K_3[B]\big)+u_{EA}}{1+K_1[A]+K_3[B]+K_1K_2[A][B]+u_{EA}+u_{EB}+u_E}
\tag{4.3}
$$

$$
p(B) = \frac{K_3[B]\big(1+K_4[B]\big)+u_{EB}}{1+K_1[A]+K_3[B]+K_1K_2[A][B]+u_{EA}+u_{EB}+u_E}
$$

$$
p(A,B) = \frac{K_1K_2[A][B]}{1+K_1[A]+K_3[B]+K_1K_2[A][B]+u_{EA}+u_{EB}+u_E}
$$

In these expressions $u_E, u_{EA}, u_{EB}, u_{EAB}$ are the drifts between standard quasi-equilibrium conditions and the actual lack of equilibrium of the system [1, 2]. It follows that the $u's$ express how the system departs from classical quasi-equilibrium conditions. Moreover, in equations (4.3) $K_1, K_2, ...$ are the ratios $k_1 / k_{-1}, k_2 / k_{-2}, ...$ of rate constants.

One can derive the expression of the "perturbation terms" u's and one finds

$$u_{EA} = \frac{kk_1[A](k_{-3} + k_4[A])}{k_{-3}k_{-4}(k_{-1} + k_2[B]) + k_{-1}k_{-2}(k_{-3} + k_4[A])}$$

$$u_{EB} = \frac{kk_3[B](k_{-1} + k_2[B])}{k_{-3}k_{-4}(k_{-1} + k_2[B]) + k_{-1}k_{-2}(k_{-3} + k_4[A])} \qquad (4.4)$$

$$u_E = \frac{k(k_{-1} + k_2[B])(k_{-3} + k_4[A])}{k_{-3}k_{-4}(k_{-1} + k_2[B]) + k_{-1}k_{-2}(k_{-3} + k_4[A])}$$

The larger the values of the perturbation terms u_{EA}, u_{EB}, u_E and the greater is the drift of the system from quasi-equilibrium.

In the case of a system such as (4.1) one can express how the binding of either substrate to the enzyme alters the mutual information of the overall process. One has

$$i(A : B) = \log \frac{p(A|B)}{p(A)} = \log \frac{p(B|A)}{p(B)} \qquad (4.5)$$

According to this expression, the mutual information will be negative if $p(A|B) > p(A)$ and positive otherwise. One can derive the expression of $p(A|B)$ and one finds

$$p(A|B) = \frac{K_4[A]}{1 + K_4[A] + u_{EB}^*} \qquad (4.6)$$

where $u_{EB}^* = u_{EB} / K_3[B]$. Hence it follows that

$$u_{EB}^* = \frac{kk_{-3}(k_{-1} + k_2[B])}{k_{-3}k_{-4}(k_{-1} + k_2[B]) + k_{-1}k_{-2}(k_{-3} + k_4[A])} \qquad (4.7)$$

and the ratio $p(A|B) / p(A)$ can be written as

$$\frac{p(A|B)}{p(A)} = \frac{K_4[A] + K_1 K_4[A]^2 + K_3 K_4[A][B] + K_1 K_2 K_4[A]^2[B] + N(u)}{K_1[A] + K_1 K_4[A]^2 + K_1 K_3[A][B] + K_1 K_3 K_4[A]^2[B] + D(u)}$$ (4.8)

Here $N(u)$ and $D(u)$ express how non-equilibrium conditions play a part in the possible existence of emergence, or in the lack of emergence, of the system. If, on the other hand, the system is under quasi-equilibrium conditions then $N(u) = D(u) = 0$. Under these conditions, emergence will be obtained if $K_1 > K_4$ and $K_3 > K_2$. If these conditions are reversed, the system can be defined as integrated. However, under conditions of quasi-equilibrium the extent of emergence, or of integration, is, of necessity, limited.

Hence it appears that the extent of emergence under non-equilibrium conditions is, to a large extent, defined by the difference $D(u)$-$N(u)$ that can be written under two equivalent forms

$$D(u) - N(u) = \left(1 - \frac{K_4}{K_1} \frac{k_{-1} + k_2[B]}{k_{-1}}\right) u_{EA} + K_1[A] u_{EB}^* + u_{EA} u_{EB}^*$$ (4.9)

$$D(u) - N(u) = u_{EA} + K_1[A]\left(1 - \frac{K_2}{K_3} \frac{k_{-3} + k_4[A]}{k_{-3}}\right) u_{EB}^* + u_{EA} u_{EB}^*$$ (4.10)

Expression (4.9) reaches maximum positive values if

$$K_1 \gg K_4 \text{ and } k_{-1} \gg k_2[B]$$ (4.11)

Similarly, expression (4.10) reaches large positive values if

$$K_3 \gg K_2 \text{ and } k_{-3} \gg k_4[A]$$ (4.12)

Under any of these conditions equations (4.9) and (4.10) reduce to

$$D(u) - N(u) = u_{EA} + K_1[A] u_{EB}^* + u_{EA} u_{EB}^*$$ (4.13)

that can only be positive. It follows that a drift from quasi-equilibrium tends to generate emergence of information. Alternatively, if conditions (4.11) and (4.12) are reversed the expression of $D(u) - N(u)$ becomes

$$D(u) - N(u) = u_{EA}\left(1 + u_{EB}^* - \frac{K_4}{K_1}\frac{k_2[B]}{k_{-1}}\right) \qquad (4.14)$$

that is likely to adopt negative values for K_1 is now much smaller than K_4 and $k_2[B]$ much larger than k_{-1}. It then follows that the system cannot be emergent.

In the case of a linear system that involves the sequential binding of substrates to the enzyme, the probabilities of occurrence of $p(A), p(B)$ and $p(A, B)$ are

$$p(A) = \frac{K_1[A] + K_1 K_2[A][B] + u_{EA}}{1 + K_1[A] + K_1 K_2[A][B] + u_{EA} + u_E} \qquad (4.15)$$

$$p(B) = p(A, B) = \frac{K_1 K_2[A][B]}{1 + K_1[A] + K_1 K_2[A][B] + u_{EA} + u_E} \qquad (4.16)$$

Similarly the conditional probability that the enzyme binds B given it has already bound A is

$$p(B|A) = \frac{K_2[B]}{1 + K_2[B] + u_{EA}^*} \qquad (4.17)$$

In expressions (4.15)-(4.17) one has

$$u_E = \frac{k}{k_{-2}} + \frac{kK_2[B]}{k_{-1}}$$

$$u_{EA} = \frac{k}{k_{-2}} + K_1[A] \qquad (4.18)$$

$$u_{EA}^* = \frac{k}{k_{-2}}$$

It follows that

$$\frac{p(B|A)}{p(B)} = \frac{1 + K_1[A] + K_1K_2[A][B] + u_{EA} + u_E}{K_1[A] + K_1K_2[A][B] + K_1[A]u_{EA}^*} \tag{4.19}$$

If the u's are close to zero, *i.e.* if the system is closed to equilibrium, one has

$$\frac{p(B|A)}{p(B)} = \frac{1 + K_1[A] + K_1K_2[A][B]}{K_1[A] + K_1K_2[A][B]} > 1 \tag{4.20}$$

Far from equilibrium one has

$$D(u) - N(u) = -\frac{k}{k_{-2}} - \frac{kK_2[B]}{k_{-1}} < 0 \tag{4.21}$$

and no emergence process is to be expected under these conditions.

2- THERMODYNAMIC IMPLICATIONS OF A DRIFT FROM QUASIEQUILIBRIUM IN A RANDOM PROCESS

When a chemical reaction, whether catalysed or not, takes place the whole system has to overtake a number of energy barriers. As a matter of fact, any rate constant is associated with an energy barrier $\Delta G^{\neq}$. Thus, for instance, a catalytic constant k can be expressed in thermodynamic terms as

$$k = \frac{k_B T}{h} \exp\left(-\Delta G^{\neq} / RT\right) \tag{4.22}$$

where k_B is the so-called Boltzmann constant, h the Planck constant, R the gas constant, T the absolute temperature and $\Delta G^{\neq}$ the free energy of activation. The higher this energy of activation and the lower the corresponding value of the rate constant.

It has already been pointed out that, in the case of emergence, one has

$$\begin{aligned} K_3 &\gg K_2 \; k_{-3} \gg k_4[A] \\ K_1 &\gg K_4 \; k_{-1} \gg k_2[B] \end{aligned} \tag{4.23}$$

It then follows that expressions (4.9) and (4.10) can be rewritten as

$$D(u) - N(u) = K_1[A]\frac{k}{k_{-2}+k_{-4}}\left(2+\frac{k}{k_{-2}+k_{-4}}\right) \tag{4.24}$$

Under this form it becomes evident that the emergent character will be enhanced if a substrate, A for instance, has a strong affinity for the enzyme and if the substrate desorption constants, k_{-2} and k_{-4}, are about the same order of magnitude as the catalytic constant (Fig. **1**). It follows from this situation that the enzyme-transition state, EX^*, is close to the ternary EAB state and both possess a high energy level.

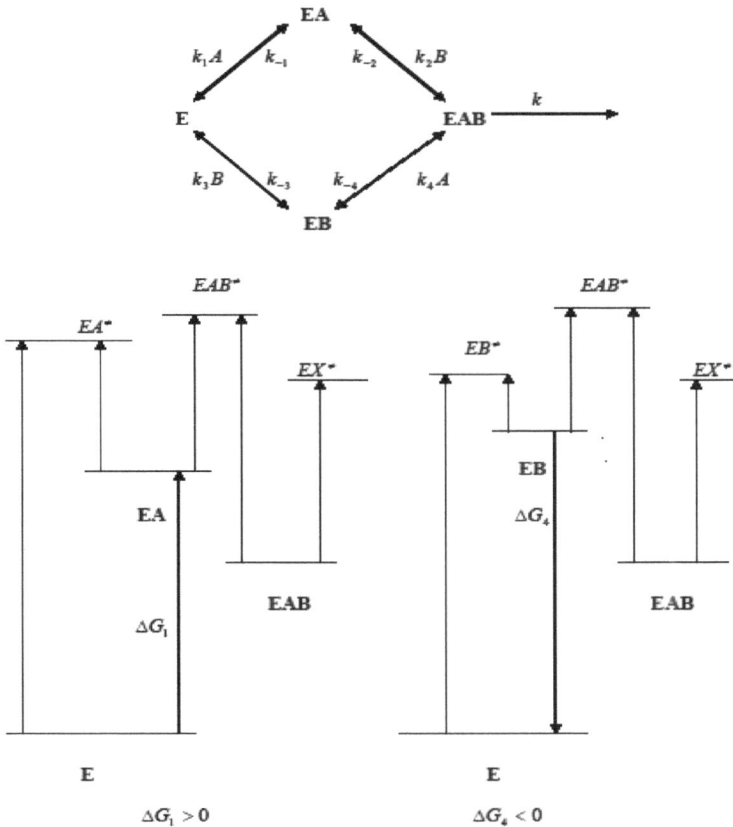

Figure 1: An integrated enzyme system. Top. Random binding of substrates A and B to the enzyme. In this Figure (heavy arrows), $\Delta G_1\rangle 0$ and $\Delta G_4\langle 0$ which means that $K_1\langle K_4$. Hence the corresponding system is integrated. Its catalytic activity is poor for the energy barrier between EAB and EX^* is high.

If the system is integrated, expression (4.24) is not valid any more. The energy profile of the reaction is now different from the previous one (Fig. **2**)

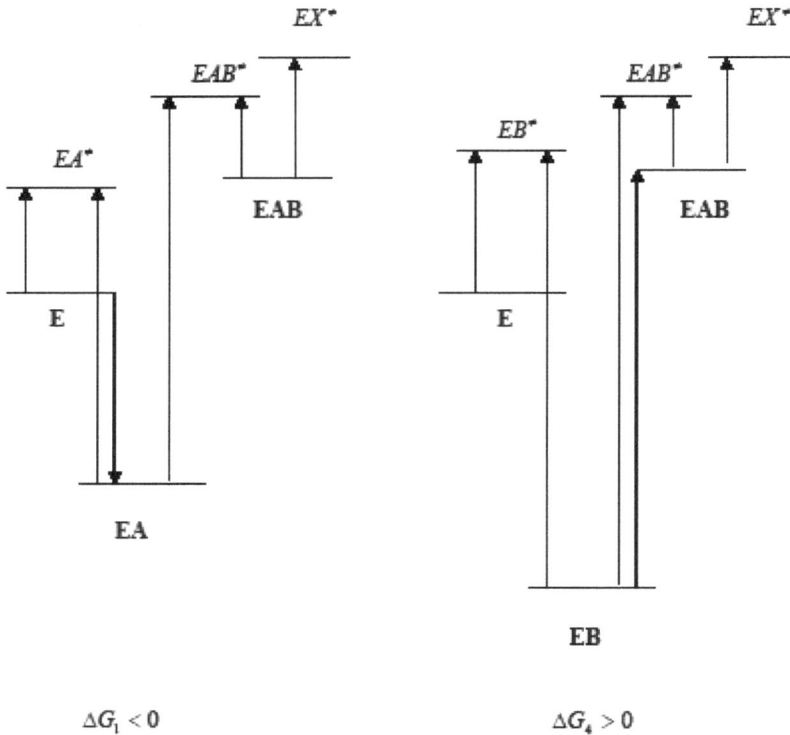

$$\Delta G_1 < 0 \qquad\qquad\qquad \Delta G_4 > 0$$

Figure 2: An emergent enzyme system. In this Fig. (heavy arrows) $\Delta G_1 < 0$ and $\Delta G_4 > 0$ which implies that $K_1 > K_4$. The corresponding enzyme system is *emergent*. Its catalytic activity is high for the energy barrier between EAB and $EX^{\neq}$ is low.

3- GENERAL CONCLUSIONS

It is often considered that the identity of living systems, whatever their complexity, is defined by the sequence of bases, or of base pairs, in macromolecules such as RNAs and DNAs. This definition of identity is by no means the only one. One can perfectly define the identity of a system without reference to a sequence of bases, or of base pairs. Under steady state, and under conditions of emergence for instance, each state of the reaction intermediates can be defined by their probabilities of occurrence. From these probabilities one can define the corresponding informations. The sequence of these informations is a good image of the identity of the nework, exactly as the sequence of nucleotides base pairs in a chromosome contributes to define the identity of a living organism.

Moreover it is remarkable that the information of any element of the reaction sequence is dependant upon all the others. For instance, the efficiency of the enzyme reaction is dependant upon the interactions present between the successive biochemical processes. For instance the information of the step $EX^{\neq}$ must be in reality defined as $h(EX^{\neq}|E, EA, EA^{\neq}, EB, EB^{\neq}, EAB, EAB^{\neq})$. It is remarkable that such a system partakes some properties with simple living systems. As simple living systems, the self-organized enzyme systems are away from equilibrium, or quasi-equilibrium, conditions and can spontaneously adjust their internal organization depending on the external milieu. The fact that such a system is away from quasi-equilibrium, can possibly give this system properties that appear novel because the lack of equilibrium may favour emergence of information in the system.

Some similarities appear to occur between the properties of simple physical-chemical systems and the way simple living organisms behave. Both possess an identity. But it is so far difficult to imagine a sensible scenario that could explain reproduction of emergent biochemical networks.

REFERENCES

[1]   Laidler K.J. (1958) The Kinetics of Enzyme Action. Oxford Clarendon Press.
[2]   Laidler K.J. and P. S. Bunting (1973) The Chemical Kinetics of Enzyme Action. Oxford Clarendon Press.
[3]   Laidler, K. J. (1969) Theories of chemical reactions rates? McGrw-Hill.
[4]   Ricard J. (2011) Biochemical lattices as models of living systems: a problem of artificial life. Trends in Cell and Molecular biology.9-29, 6, 2011.

Send Orders for Reprints to reprints@benthamscience.net

CHAPTER 5

Enzyme Activity within the Living Cell

Abstract: Enzyme processes do not display any apparent "finality" when considered isolated, however they do within the cell. In fact the global behavior of the cell relies upon a co-ordination of many enzyme reactions. The aim of the so-called "metabolic control theory" is to show how many enzyme reactions are connected as to form a coherent network that possesses properties that are quantitatively and qualitatively different from those of individual enzyme reactions. The concept of "cascade" is discussed in this Chapter and their properties are shown to be different from those of the same enzymes considered in isolation. Many enzymes within the cell are associated with cell structures that behave as polyanions. Electrostatic repulsion of negatively charged substrates by the negative charges of plant cell wall generates an apparent negative co-operativity of enzyme reactions that take place in the cell wall. These effects can be modulated by external ionic strength. It follows that the behavior of enzymes in the cell have properties quite different from the same enzymes in free solution.

Keywords: Metabolic control theory, Parameters of a metabolic process, Summation theorems, Homogeneous functions, Elasticity of an enzyme reaction, Enzyme cascades, Electrostatic partitioning of ions, Enzymes and electrostatic partitioning, Donnan equation, Bound-enzyme activity and heterogeneously charged matrices, Enzyme activity and heterogeneously charged matrices, Enzyme co-operativity and complexity of charged matrices, Enzymes and plant cell wall extension.

It is well known that the enzyme reactions that take place in the living cell are sensitive to a large number of chemical substances. Hence it has for long been concluded that the control of metabolism was exerted through the action of certain chemical substances on individual enzymes. In a way, the control of metabolism was studied in a reductionist manner for it was believed it could be considered as identical to that of individual enzymes. There are at least three reasons to reject this type of approach.

First, different enzymes take part in the same metabolic pathway and, for that reason, possess a global behaviour that cannot be explained by the sole properties of these individual enzymes. As a matter of fact, two enzymes that take place in succession in the same pathway share a common reaction intermediate in such a

Jacques Ricard

way that even the simplest metabolic process should be considered a system that possess global properties distinct from the behaviour of the enzymes present in that metabolic chain.

Second, many reactions within the cell are controlled by another reaction that catalyse the reverse process. As we shall see latter such systems usually possess global properties that are markedly different from the individual properties of the enzymes that constitute the system.

Third, many enzymes act on substrates, or interact with ligands that are in fact ions. Moreover many proteins, cell organelles, membranes are in fact charged structures that are extremely sensitive to the local ionic strength. It follows from this reasoning that the behaviour of enzyme reactions within the cell can be markedly dependant upon the ionic conditions that prevail in this heterogeneous system.

1- METABOLIC CONTROL THEORY

In contrast with the reductionist vision of metabolism that postulates the existence of "key enzymes" responsible for the global behaviour of metabolic processes the "new" theories of metabolism develop the idea that processes should be described in a global manner. The bases of metabolic control theory have been formulated by Kacser and Burns [1-5] as well as by Heinrich and Rapoport [6-9]. A different approach that will not be presented here has also been proposed by Savageau and associates [10,11].

1.1 - The Main Features of Metabolic Control Theory

Let us consider a linear sequence of enzyme reactions

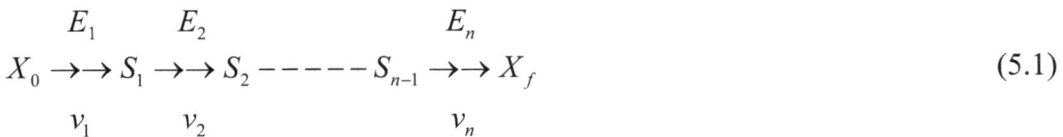

$$
\begin{array}{cccc}
E_1 & E_2 & & E_n \\
X_0 \to\to S_1 \to\to S_2 & ----- & S_{n-1} \to\to X_f \\
v_1 & v_2 & & v_n
\end{array}
\tag{5.1}
$$

Here X_0 and X_f are defined as the source and the sink of the process. By convention they are considered very large and identified to reservoirs. The intermediates of the process are $S_1, S_2, ... S_{n-1}$. Their mutual interconnection is effected thanks to the enzymes $E_1, E_2, ... E_N$. If the whole system is in steady state, the overall flux J is equal to the rate of any step. One has then

$$J = \frac{\partial S_{n-1}}{\partial t} = v_1 = v_2 = ... = v_{n-1} \qquad (5.2)$$

1.1.1- Fundamental Parameters of a Metabolic Process

One can easily define three types of parameters that control the overall flux: the flux control coefficient, the concentration control coefficient and the elasticities. The flux control coefficients are defined as

$$C_i^j = \frac{E_i}{J} \frac{\partial J}{\partial E_i} = \frac{\partial \ln J}{\partial \ln E_i}. \qquad (5.3)$$

These coefficients express how the system, as a whole, reacts to slight changes of enzyme concentrations. The second type of coefficient is called the concentration control coefficient

$$C_i^{sj} = \frac{E_i}{S_j} \frac{\partial S_j}{\partial E_i} = \frac{\partial \ln S_j}{\partial \ln E_i} \qquad (5.4)$$

This parameter expresses how a concentration change of the enzyme E_i affects, in the system, the concentrations of the intermediates S_j. Then the parameter C_i^{sj} describes a systemic property of the network. The last parameter, ε_{Sj}^{vi}, called elasticity, describes how a change of concentration of the intermediate S_j reverberates to the rate v_i of the i th step. It is defined as

$$\varepsilon_{sj}^{vi} = \frac{S_j}{v_j} \frac{\partial v_i}{\partial S_j} = \frac{\partial \ln v_i}{\partial \ln S_j} \qquad (5.5)$$

In that case, the response of the network is not systemic but affects one step only.

1.1.2- Summation Theorems

If we do not pay attention to the concentrations of reaction intermediates $S_1, S_2, ...$ the overall flux of the metabolic sequence is a function of the enzyme concentrations. One has

$$J = f(E_1, E_2, ...) \tag{5.6}$$

Moreover it is evident that this flux is a function of enzyme concentrations. One has then

$$dJ = \sum_i \left(\frac{\partial J}{\partial E_i} \right) dE_i \tag{5.7}$$

As $\partial J / \partial E_i = C_i^J J / E_i$ it follows that

$$\frac{\partial J}{j} = \sum_i C_i^J \frac{dE_i}{E_i} \tag{5.8}$$

If the perturbations dE_E / E_i are invariant and equal to α the relative change of the reaction flux will also be equal to α. Hence one has

$$\alpha = \sum_i C_i^J \alpha \tag{5.9}$$

and this implies that

$$\sum_i C_i^j = 1 \tag{5.10}$$

This is the mathematical expression of the so-called summation theorem of the flux control coefficients. This theorem states that if the steady state of a system is unchanged, despite the perturbation of the reaction flux, the control coefficients cannot be independent and their sum must be equal to one. This conclusion is clearly at variance with the classical assumption commonly made that metabolic pathways are controlled through specific enzymes sensitive to an end-product of the reaction sequence.

In fact, the summation theorem is equivalent to the statement that a flux J is a homogeneous function of degree one in enzyme concentrations $E_1, E_2, ...$ A function $f(x_1, x_2, ...)$ is called homogeneous of degree h if

$$f(tx_1, tx_2, ...) = t^h f(x_1, x_2, ...) \tag{5.11}$$

Euler's theorem expresses the view that a function $f(x_1, x_2, ...)$ is homogeneous of degree h in $x_1, x_2, ...$ if

$$hf(x_1, x_2, ...) = \sum_j x_j \frac{\partial f}{\partial x_j} \tag{5.12}$$

Conversely any function $g(x_1, x_2, ...)$ that obeys the following relationship

$$hg(x_1, x_2, ...) = \sum_j x_j \frac{\partial g}{\partial x_j} \tag{5.13}$$

should be homogeneous of degree h in $x_1, x_2, ...$ The general expression of the summation theorem is then

$$J = \sum_j \left(\frac{\partial J}{\partial E_j} \right) E_j \tag{5.14}$$

The immediate consequence of this result is a simple relationship between the behaviour of the chain and the individual enzyme reactions. If applied to a step j the relationship above becomes

$$\left(\frac{\partial v_j}{\partial E_j} \right) E_j = v_j \tag{5.15}$$

thus implying that

$$tv_j(E_j) = v_j(tE_j) \tag{5.16}$$

This relationship means that changing the enzyme concentration by a value t leads to an increase of the rate equal to the same value.

One may also state that any substrate concentration S_j is a function of the set of enzyme concentrations. One has

$$S_j = g(E_1, E_2, ...) \tag{5.17}$$

Perturbing these enzyme concentrations by the following values, $dE_1, dE_2,$ ....generates a perturbation of the substrate concentration, dS_j, and one has

$$dS_j = \sum_i \left(\frac{\partial S_j}{\partial E_i}\right) dE_i \tag{5.18}$$

Dividing by S_j both sides of this expression and taking advantage of the definition of substrate control coefficient leads to

$$\frac{dS_i}{S_j} = \sum_i C_i^{Sj} \frac{dE_i}{E_i} \tag{5.19}$$

Moreover perturbing the enzyme concentration by the same value, α, will not affect the steady state concentration S_j which means in turn that

$$\sum_i C_i^{Sj} \alpha = 0 \tag{5.20}$$

which requires that

$$\sum_i C_i^{Sj} = 0 \tag{5.21}$$

1.1.3- Elasticities

Let us consider a given step, i, in a metabolic network. Its steady state rate, v_i, is a function of both the corresponding substrates and enzyme concentration, *i.e.* $S_i, ..., S_j$ and E_i, respectively. One has

$$v_i = f(E_i, S_i, ..., S_j) \tag{5.22}$$

The total differential of the velocity v_i is then

$$dv_i = \left(\frac{\partial v_i}{\partial E_i}\right) dE_i + \sum_j \left(\frac{\partial v_i}{\partial S_j}\right) dS_j \tag{5.23}$$

This expression may then be rewritten as

$$dv_i = \frac{\partial v_i}{\partial E_i} \frac{E_i}{v_i} \frac{dE_i}{E_i} v_i + \sum_j \frac{\partial v_i}{\partial S_j} \frac{S_j}{v_i} \frac{dS_j}{S_j} v_i \qquad (5.24)$$

If the enzyme E_i does not interact with other enzymes, then

$$\frac{\partial v_i}{\partial E_i} \frac{E_i}{v_i} = 1 \qquad (5.25)$$

and taking advantage of equation (5.5) one can rewrite equation (5.24) as

$$dv_i = \frac{dE_i}{E_i} v_i + \sum_j \varepsilon_{Sj}^{vi} \frac{dS_j}{S_j} v_i \qquad (5.26)$$

The ε parameter can be considered a "term of elasticity" as it shows that, if the concentration of enzyme E_i is perturbed, the corresponding rate, v_i, remains unchanged provided that

$$\frac{dE_i}{E_i} = -\sum_j \varepsilon_{Sj}^{vi} \frac{dS_j}{S_j} \qquad (5.27)$$

This expression defines the condition required to maintain the flux constant when the other intermediates of the pathway are perturbed by the values $dS_1, dS_2, \ldots$.

Without going too far in the intricacies of the present theory, it appears that it is incompatible with the classical reductionist approach of metabolic pathways regulation. As it has been shown in the present Section, a metabolic pathway is considered a global structure that may change and adapt its properties to the variations of substrate concentrations. Its global properties cannot be understood by the sole study of one of its protein components.

2- ENZYME CASCADES

Many biochemical processes share an apparently surprising mechanism. This mechanism is shown in Fig. **1**. During this process, a protein P is being converted into a modified form P^* thanks to an enzyme E_1 that plays the part of a converter. The modified protein P^* is then converted back to P thanks to another enzyme E_2.

In this perspective, the conversion $P \rightarrow P^*$ could be a methylation or a phosphorylation process. The cyclic process shown in Fig. **1** is called a monocyclic cascade [12-21].

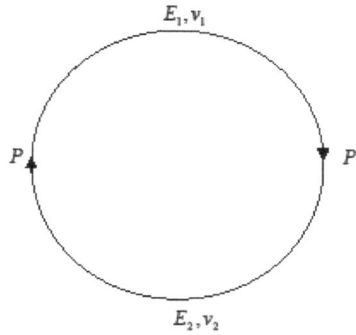

Figure 1: A monocyclic cascade. A protein P is converted into a modified form and back thanks to two enzymes E_1 and E_2. v_1 and v_2 are the rates of conversion of P into P^*, and back.

The conservation equation for protein P is

$$P_T = P + P^* \tag{5.28}$$

Where P_T is the total protein concentration. If we assume that the enzyme process follows classical Michaelis-Menten kinetics one has

$$v_1 = \frac{V_1\left(P_T - P^*\right)}{K_1 + \left(P_T - P^*\right)} \tag{5.29}$$

and

$$v_2 = \frac{V_2 P^*}{K_2 + P^*} \tag{5.30}$$

V_1 and V_2 are the maximum rates, K_1 and K_2 the Michaelis constants. Setting

$$\tilde{K}_1 = \frac{K_1}{P_T} \quad \tilde{K}_2 = \frac{K_2}{P_T} \quad \tilde{P}^* = \frac{P^*}{P_T} \quad \alpha = \frac{V_1}{V_2} \tag{5.31}$$

Equations (5.29) and (5.30) reduce to

$$\tilde{v}_1 = \frac{V_1\left(1-\tilde{P}^*\right)}{\tilde{K}_1+\left(1-\tilde{P}^*\right)} \tag{5.32}$$

$$\tilde{v}_2 = \frac{V_2\tilde{P}^*}{\tilde{K}_2+\tilde{P}^*} \tag{5.33}$$

As these two reactions are coupled one has

$$\frac{V_1\left(1-\tilde{P}^*\right)}{\tilde{K}_1+\left(1+\tilde{P}^*\right)} = \frac{V_2\tilde{P}^*}{\tilde{K}_2+\tilde{P}^*} \tag{5.34}$$

that can be rearranged to

$$(\alpha-1)\tilde{P}^{*2} - \left\{(\alpha-1) - \tilde{K}_2\left(\alpha+\frac{\tilde{K}_1}{\tilde{K}_2}\right)\right\}\tilde{P}^* - \alpha\tilde{K}_2 = 0 \tag{5.35}$$

The variation of $\tilde{P}^*$ as a function of α can display an all-or-none type of behaviour if α is close to unity (Fig. **2**).

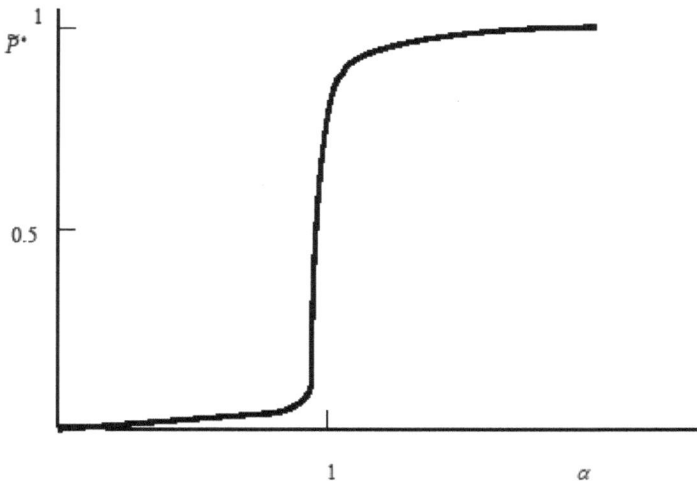

Figure 2: All-or-none type of variation of $\tilde{P}^*$ as a function of α . See text.

Let us consider now two converter enzymes of a monocyclic cascade and let's assume these enzymes to be sensitive, to a different extent to the same ligand L (Fig. **3**). The amplification factor A_L of the response to a change of effector concentration can be defined as

$$A_L = \left(\frac{\partial P^*}{\partial L} \right) \frac{L}{P^*} = \frac{\partial \ln P^*}{\partial \ln L} \tag{5.36}$$

As shown in equation (5.35) and in the corresponding Fig. **2**, P^* must be a function of α and α must be a function of L

$$\tilde{P}^* = F(\alpha) \tag{5.37}$$

$$\alpha = g(L)$$

The usual rule of differentiation of a function of function allows one to write

$$\frac{\partial \tilde{P}^*}{\partial L} = \frac{\partial \tilde{P}^*}{\partial \alpha} \frac{\partial \alpha}{\partial L} \tag{5.38}$$

and

$$A_L = \left(\frac{\partial \tilde{P}^*}{\partial L} \right) \frac{L}{T^*} = \left(\frac{\partial \tilde{P}^*}{\partial \alpha} \right) \frac{\alpha}{\tilde{P}^*} \left(\frac{\partial \alpha}{\partial L} \right) \frac{L}{\alpha} \tag{5.39}$$

The first factor, $\left(\partial \tilde{P}^* / \partial \alpha \right)\left(\alpha / \tilde{P}^* \right)$ measures the steepness of the curves such as those shown in Fig. **2**. The second factor $\left(\partial \alpha / \partial L \right)\left(L / \alpha \right)$ enhances the amplification of the response if it assumes values larger than unity. Hence it is of interest to derive the expression of this factor.

If one considers the converter enzyme system of Fig. **3**, the functions, f_1 and f_2, of the converter enzymes are

$$f_1 = \frac{K_1'[L]}{1 + K_1'[L]} \tag{5.40}$$

$$f_2 = \frac{1}{1+K_2'[L]}$$

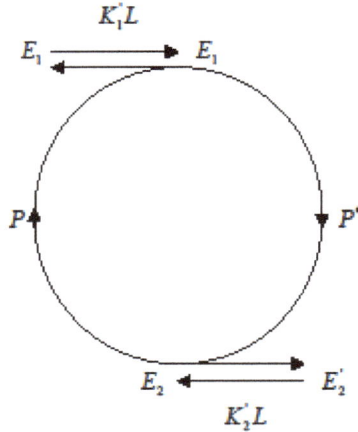

Figure 3: Action of a ligand L on a monocyclic cascade. See text.

where K_1' and K_2' are the relevant equilibrium constants. It follows that the two maximum rates, V_1 and V_2, associated with the conversion processes are expressed as

$$V_1 = k_1[E_1]_T = \frac{k_1[E_1]_T K_1'[L]}{1+K_1'[L]} \tag{5.41}$$

$$V_2 = k_2[E_2]_T = \frac{k_2[E_2]_T}{1+K_2'[L]}$$

where k_1 and k_2 are the rate constants of the two antagonistic processes, $[E_1]_T$ and $[E_2]_T$ are the total concentrations of the converter enzymes. It follows that the expression of α is

$$\alpha = \frac{k_1[E_1]_T}{k_2[E_2]_T} \frac{K_1'[L]+K_1'K_2'[L]^2}{1+K_1'[L]} \tag{5.42}$$

Its derivative assumes the form

$$\frac{\partial \alpha}{\partial [L]} = \frac{k_1 [E_1]_T}{k_2 [E_2]_T} \frac{K_1' \left(1 + 2K_2' [L] + K_1' K_2' [L]^2\right)}{\left(1 + K_1'\right)^2} \tag{5.43}$$

Hence one has

$$\frac{\partial \alpha}{\partial [L]} \frac{[L]}{\alpha} = \frac{1 + 2K_2' [L] + K_1' K_2' [L]^2}{1 + \left(K_1' + K_2'\right)[L] + K_1' K_2' [L]^2} \tag{5.44}$$

One can notice that this expression will be larger than one if $K_1' < K_2'$. In that case the effector will enhance the amplification of the response.

Many cascades are polycyclic. An example of a bicyclic cascade is shown in Fig. **4**. In this simple bicyclic cascade it is assumed that a protein, P^*, plays the part of a converter enzyme involved in the conversion $Q \rightarrow Q^*$. In the case of this bicyclic cascade the enhancement of the response will be obtained if

$$\left(\frac{\partial \tilde{Q}^*}{\partial \alpha}\right) \frac{\alpha}{\tilde{Q}^*} > \left(\frac{\partial \tilde{P}^*}{\partial \alpha}\right) \frac{\alpha}{\tilde{P}^*} \tag{5.45}$$

The maximum reaction rates of the two cascades are V_1 and V_2 for the first, V_3 and V_4 for the second (Fig. **4**). The variables α and α' for the two cascades are

$$\alpha = \frac{V_1}{V_2} = \frac{k_1 [E_1]}{k_2 [E_2]} \tag{5.46}$$

$$\alpha' = \frac{V_3}{V_4} = \frac{k_3 P^*}{k_4 [E_4]}$$

Moreover, $\tilde{P}^*$ is a function of α and $\tilde{Q}^*$ is a function of α'. One has thus

$$\tilde{P}^* = f(\alpha) \tag{5.47}$$

$$\tilde{Q}^* = g(\alpha')$$

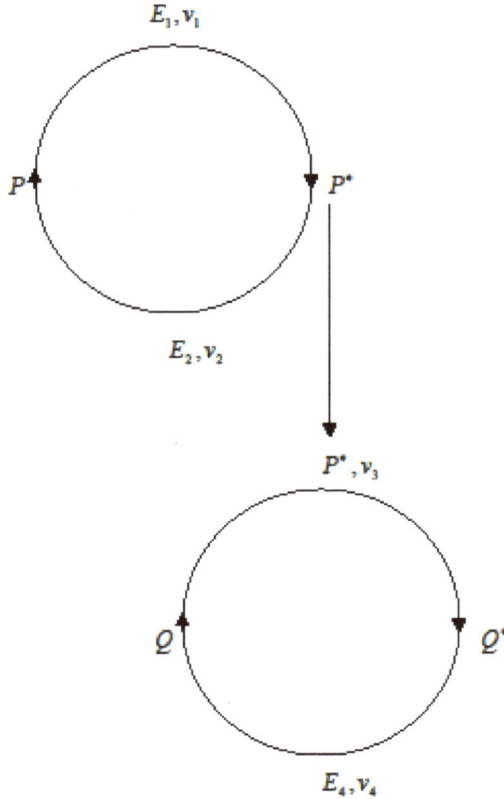

Figure 4: A bicyclic cascade. As previously, v_1, v_2, v_3, v_4 correspond to reactions rates of conversion. See text.

The functions of α and α' can be derived by solving two quadratic equations similar to equation (5.35). The rule of differentiation of functions of functions allows one to write

$$\left(\frac{\partial \tilde{Q}^*}{\partial \alpha}\right)\frac{\alpha}{\tilde{Q}^*} = \left(\frac{\partial \tilde{Q}^*}{\partial \alpha'}\right)\frac{\alpha'}{\tilde{Q}^*}\left(\frac{\partial \alpha'}{\partial \tilde{P}^*}\right)\frac{\tilde{P}^*}{\alpha'}\left(\frac{\partial \tilde{P}^*}{\partial \alpha}\right)\frac{\alpha}{\tilde{P}^*} \tag{5.48}$$

This expression shows that changing the enzyme ratio of the first cascade interacts with the behaviour of the second one.

An important point that should be stressed is that the efficiency of a bicyclic cascade relies upon the coupling of the two elementary cascades *viz.*

$$\left(\frac{\partial \tilde{Q}^*}{\partial \alpha'}\right)\frac{\alpha'}{\tilde{Q}^*}\left(\frac{\partial \alpha'}{\partial \tilde{P}^*}\right)\frac{\tilde{P}^*}{\alpha'} \tag{5.49}$$

If this expression is larger than one, the whole system will have its efficiency increased owing to the existence of the bicyclic cascade.

Hence it appears that an enzyme in a cell does not act in isolation is such a way it may appear nonsense to reduce the chemistry of the living cell to the individual functioning of independent enzymes.

3- ENZYME REACTIONS WITHIN THE LIVING CELLS AND ELECTROSTATIC PARTITIONING

The cell milieu of a living organism constitutes an extremely complex system. This system is made up of macromlecules and macromolecular aggregates that behave as polycations or polyanions. Plant cell walls and ribosomes constitute typical examples of these charged biological structures. Moreover enzyme reactions that take place in living cells often involve ions as substrates, or receptors. It is then evident that if an enzyme reaction, involving charged substrates, takes place at the surface of the cell, the kinetics and the properties of this reaction will be affected owing to the phenomenon of charge-charge interactions. As we shall see, such a situation generates unexpected kinetic behaviour that is not observed with purified enzymes in free solution [22-35].

3.1 - Charged Matrices and Electrostatic Partitioning of Ions

Let us consider mobile ions within a polyelectrolyte matrix. Their electrochemical potential can be expressed as

$$\tilde{\mu}_i = \mu^0 + RT \ln(\gamma_i c_i) + zF\psi_i \tag{5.50}$$

In this expression $\tilde{\mu}_i$ is the electrochemical potential of the ion. γ_i and c_i are the corresponding activity coefficient and concentration, respectively. ψ_i and z are the electrostatic potential and the valence of the ion multiplied by $+1$ or -1 depending on the ion is a cation or an anion. μ^0 is the standard chemical potential of the ion. R and T have their usual significance. The subscript i refers to the inside of the matrix. Similarly the electrochemical potential of the same ion outside the matrix is

$$\tilde{\mu}_0 = \mu^0 + RT \ln(\gamma_0 c_0) + zF\psi_0 \tag{5.51}$$

Here the index "0" refers to the outside of the charged matrix. Under thermodynamic equilibrium conditions one should have

$$\tilde{\mu}_0 - \tilde{\mu}_i = 0 = RT \ln \frac{\gamma_0 c_0}{\gamma_i c_i} + zF(\psi_0 - \psi_i) \tag{5.52}$$

This expression can be rearranged to

$$\frac{1}{z} \ln \frac{\gamma_i c_i}{\gamma_0 c_0} = \frac{F\Delta\Psi}{RT} \tag{5.53}$$

where $\Delta\Psi$ is defined as

$$\Delta\Psi = \psi_0 - \psi_i \tag{5.54}$$

Expression (5.53) is equivalent to a Nernst equation. If a cation, A^{z+}, and an anion, B^{z-} (with $z+ = z_A$ and $z- = -z_B$) are present in the two phases one has

$$\frac{1}{z_A} \ln \frac{\gamma_i^A C_i^A}{\gamma_0^A C_0^A} = \frac{1}{z_B} \ln \frac{\gamma_0^B C_0^B}{\gamma_i^B C_i^B} = \frac{F\Delta\Psi}{RT} \tag{5.55}$$

where γ^A, γ^B and C^A, C^B represent the activity coefficients and concentrations of the two ions. One can define the electrostatic coefficient Π as

$$\Pi_e = \exp\left(\frac{F\Delta\Psi}{RT}\right) \tag{5.56}$$

and equation (5.55) above becomes

$$\left(\frac{\gamma_i^A C_i^A}{\gamma_0^A C_0^A}\right)^{1/z_A} = \left(\frac{\gamma_0^B C_0^B}{\gamma_i^B C_i^B}\right)^{1/z_B} = \Pi_e \tag{5.57}$$

which is the well-known Donnan equation.

Let us consider the situation where the inside and outside of the polyelectrolyte matrix contain different anions with different concentrations and one type only of cation. Let us call $B^-, B^{2-}, ..., B^{z-}$ and A^+ these anions and cation. One has then

$$\sum B_0^- + ... + z\sum B_0^{z-} = \sum A_0^+ \qquad (5.58)$$

$$\sum B_i^- + ... + z\sum B_i^{z-} + \Delta^- = \sum A_i^+$$

In the second equation above Δ^- is the fixed negative charge density of the matrix. This equation can be rewritten as

$$\frac{\sum B_0^-}{\Pi_e} + ... + \frac{z\sum B_0^{z-}}{\Pi_e^z} + \Delta^- = \Pi_e \sum A_0^+ \qquad (5.59)$$

Let us consider for instance the simple situation where all the anions in the matrix are monovalent and where the matrix itself is a polyanion. One has then

$$\Pi_e = \frac{\sum B_0^-}{\sum B_i^-} = \frac{\sum A_i^+}{\sum A_0^+} \qquad (5.60)$$

The two electroneutrality conditions imply that

$$\sum B_0^- = \sum A_0^+ \qquad (5.61)$$

$$\sum B_i^- + \Delta^- = \sum A_i^+$$

The second equation above can be rewritten as

$$\Pi_e^2 - \frac{\Delta^-}{\sum B_o^-}\Pi_e - 1 = 0 \qquad (5.62)$$

and its positive root assumes the form

$$\Pi_e = \frac{\Delta^-}{2\sum B_0^-} + \frac{1}{2}\sqrt{\left(\frac{\Delta^-}{\sum B_0^-}\right)^2 + 4} \tag{5.63}$$

If the matrix were a polycation one would have obtained

$$\Pi_e = \frac{1}{2}\sqrt{\left(\frac{\Delta^+}{\sum B_0^-}\right)^2 + 4} - \frac{\Delta^+}{2\sum B_0^-} \tag{5.64}$$

For the simple situation that involves an insoluble polyanion, a monovalent anion B^- and a cation A^+ the partition coefficient decreases as the anion concentration increases (Fig. **5**).

The immediate consequence of these results is a change of the pH-profile of the reaction. Thus, for instance, if the substrate is uncharged and if the polyelectrolyte is a polyanion, the pH-profile is shifted towards high values by a value equal to $\log \Pi_e$. If the substrate is itself an anion, the pH-profile is even more complex. Under these conditions the plot of $\log\left(k_c^* / \tilde{K}_D^*\right)$ (where k_c^* is the apparent catalytic constant and $\tilde{K}_D^*$ the apparent dissociation constant) *versus* pH is shifted by a value of $2\log \Pi_e$.

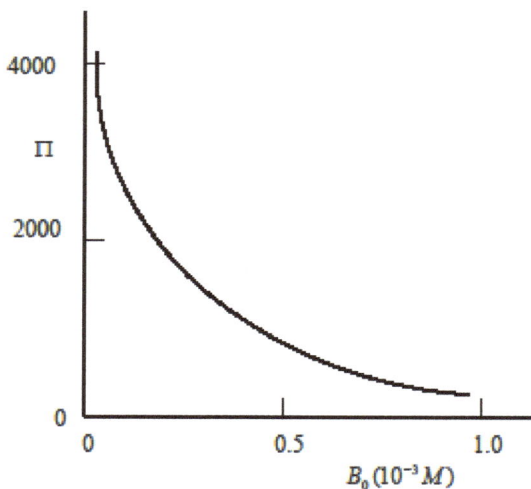

Figure 5: Variation of the electrostatic partition coefficient Π as a function of the bulk anion concentration B_0. See text.

3.2- Charged Matrices and Apparent Kinetic Co-Operativity of Polyelectrolyte-Bound Enzymes

We have considered so far the pH-profile shifts when the substrate, in addition to the matrix, is negatively charged. The electrostatic interaction effects, however, may induce apparent co-operativity of the bound-enzyme reaction. Let us consider the simple process

$$v = \frac{V_m[S_i]}{K_n + [S_i]} \tag{5.65}$$

where V_m and K_m are maximum reaction velocity and the Michaelis constant of the reaction, respectively. As usually, $[S_i]$ is the local concentration of a monovalent substrate. Under these conditions one has

$$\Pi_e = \frac{[S_0]}{[S_i]} \tag{5.66}$$

and it follows that

$$[S_i] = \frac{[S_0]}{\Pi_e} \tag{5.67}$$

and equation (5.65) becomes

$$v = \frac{V_m[S_0]/\Pi_e}{K_m + [S_0]/\Pi_e} \tag{5.68}$$

Now, if the fixed charge density Δ^- is much larger than the charge density brought about by the substrate, then (equation 5.63)

$$\left(\frac{\Delta^-}{\sum S_0}\right)^2 \gg 4 \tag{5.69}$$

and equation (5.63) reduces to

$$\Pi_e = \frac{\Delta^-}{[S_0]} \tag{5.70}$$

It follows that equation (5.68) reduces to

$$v = \frac{V_m [S_0]^2}{K_m \Delta^- + [S_0]^2} \tag{5.71}$$

As a consequence, a plot of v as a function of $[S_0]$ will result in a sigmoidal curve that does not express any real co-operativity. In fact positive co-operativity is only apparent and originates from the electrostatic repulsion effects between the fixed charges of the matrix and the charges of the substrate. Last but not least, the sigmoidal behaviour will become hyperbolic if v is plotted against $\sigma_0 = [S_0]^2 / K_M$.

3.3- Heterogeneously Charged Matrices and Enzyme Activity

It is highly probable that the plant cell wall is not a homogeneous medium *viz.* the fixed negative charges and the enzyme molecules are not homogeneously distributed within the cell wall. Hence one can wonder about the effect of charge and enzyme distribution on the overall enzyme activity. In order to answer this question it is important to know whether it is possible to express quantitatively the degree of order that may, or may not, exist in the spatial distribution of fixed charges and bound enzyme molecules. It is also important to understand what is meant by the term organization. In the following, this term is taken to mean the lack of pure randomness of fixed charges and bound enzyme molecules.

One can express quantitatively the degree of organization of such a medium by the monovariate moments of charge and enzyme distribution as well as by the bivariate moments that associate these two distributions. One can express, in particular, the local charge density, Δ_i, of the charged clusters with respect to their mean $<\Delta>$ and similarly the maximum reaction velocity, V_j, which is proportional to local enzyme density, relative to its mean $<V>$. One has

$$\Delta_i = <\Delta> + \delta_i \quad (i = 1,....,n) \tag{5.72}$$

$$V_j = <V> + \varepsilon_j \quad (j = 1, ..., n)$$

where δ_i and ε_j are the individual deviations relative to their corresponding mean. The distributions of δ_i and ε_j may be mono- or multivatiate and may be characterized by their moments. Monovariate moments take account of one variable only and are defined as

$$\sum_i f_i \delta_i = N\mu_1(\delta) = 0$$

$$\sum_j f_j \varepsilon_j = N\mu_1(\varepsilon) = 0$$

$$\sum_i f_i \delta_i^2 = N\mu_2(\delta) = 0 \qquad (5.73)$$

$$\sum_j f_j \varepsilon_j^2 = N\mu_2(\varepsilon) = 0$$

$$\sum_i f_i \delta_i^3 = N\mu_3(\delta)$$

$$\sum_j f_j \varepsilon_j^3 = N\mu_3(\varepsilon)$$

In these expressions, $\mu_1(\delta), \mu_2(\delta), \mu_1(\varepsilon)$ and $\mu_2(\varepsilon)$ are the monovariate and bivariate moments of charge density and reaction velocity, respectively. Multivariate moments are also required to characterize enzyme and charge density distributions. For instance, the bivariate moments are defined as

$$\sum_i \sum_j f_{ij} \delta_i \varepsilon_j = N\mu_{11}(\delta, \varepsilon)$$

$$\sum_i \sum_j f_{ij} \delta_i^2 \varepsilon_j = N\mu_{21}(\delta, \varepsilon) \qquad (5.74)$$

$$\sum_i \sum_j f_{ij} \delta_i^3 \varepsilon_j = N\mu_{31}(\delta, \varepsilon)$$

In these expressions, the three $\mu(\delta,\varepsilon)$ represent the bivariate moments.

These general principles can be applied to different types of organization of fixed charges and enzyme molecules in the cell wall (Fig. **6**). If the charge densities and enzyme molecules are independently distributed (Fig. **6A**) then $\mu(\delta)=\mu(\varepsilon)=\mu(\delta,\varepsilon)=0$. Both the charges and the enzyme molecules are randomly distributed over the surface. In the second type of organization (Fig. **6B**) the charges are organized in space but the enzyme molecules are not. One should then expect some moments $\mu(\delta)$ to be different from zero whereas all the moments $\mu(\varepsilon)$ should be equal to zero. It then follows that all the bivariate moments $\mu(\delta,\varepsilon)=0$. In the third type of organization (Fig. **6C**) the enzyme molecules are spatially organized but the fixed charges are not. Hence some moments $\mu(\varepsilon)$ should be different from zero whereas all the moments $\mu(\delta)$ should be equal to zero. As a consequence, all the bivariate moments should be nil.

Following this reasoning allows understanding how the moments of the distribution previously defined express, in quantitative terms, the different possible types of cell wall organization. In the case of Fig. **6A**, there is no spatial organization of fixed charges relative to enzyme molecules for the monovariate $\mu(\delta)$ and $\mu(\varepsilon)$ as well as the bivariate $\mu(\delta,\varepsilon)$ are all equal to zero. In the second type of organization (Fig. **6B**), some monovariate moments $\mu(\delta)$ are different from zero whereas the monovariate and bivariate moments, $\mu(\varepsilon)$ and $\mu(\delta,\varepsilon)$, are nil. This means the existence of an organization for the enzyme molecules, but not for the fixed charges. In the third type of organization (Fig. **6C**) $\mu(\varepsilon)$ is different from zero, which implies the existence of some type of organization for the enzyme molecules. However, in this case $\mu(\delta)$ and $\mu(\delta,\varepsilon)$ are equal to zero which means a lack of spatial organization for fixed charges. In the D type of organization, one should expect that some monovariate and bivariate moments, $\mu(\delta)$, $\mu(\varepsilon)$ and $\mu(\delta,\varepsilon)$ be different from zero. Last, in the E type of organization, clusters of fixed charges and enzyme molecules are superimposed and the monovariate and bivariate moments should be equal to zero. Hence one can conclude that when monovariate and bivariate moments are different from zero then the clusters of enzyme molecules and charges should be superimposed (Fig. **6D**).

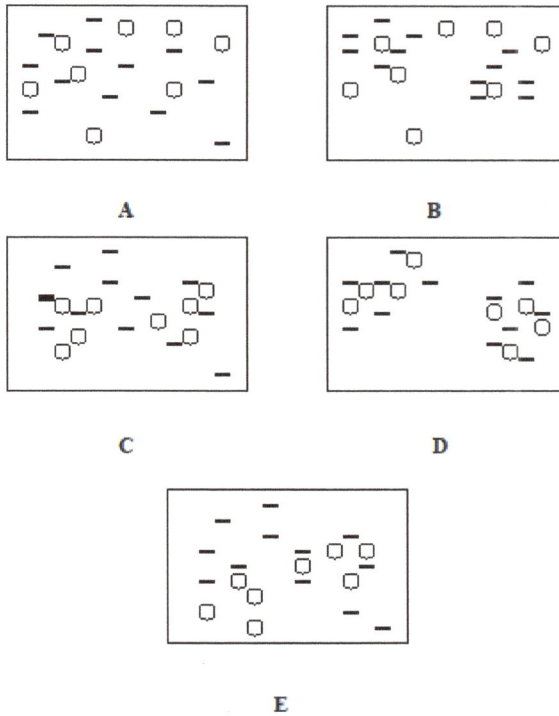

Figure 6: Different possible types of enzyme and negative charge distribution. A-Fixed charges and enzyme molecules are randomly distributed. B-Fixed charges are clustered but enzyme molecules are randomly distributed. C and E-Enzyme molecules are clustered but fixed charges are randomly distributed. D-Enzyme molecules and charges are clustered mutualy overlap.

3.4- Apparent Co-Operativity and Complexity of the Polyelectrolyte Matrix

It was shown in the previous Section that the polyelectrolyte matrix has a complex structure when the monovariate and bivariate moments associated with charge and enzyme densities are not equal to zero. If, for instance, the substrate is a monoanion, as shown in Fig. **6E**, the enzyme reaction rate will be

$$v = \sum_i \sum_j \frac{f_{ij} V_j S_0}{K \Pi_i + S_0} \qquad (5.75)$$

In this expression, Π_i is the partition coefficient of charge cluster i and V_j the maximum velocity proportional to the enzyme density present in cluster j. If, for simplicity, the substrate is a monoanion, expression (5.75) assumes the form

$$v = \sum_i \sum_j \frac{f_{ij} V_j S_0^2}{K \Delta_i + S_0^2} \tag{5.76}$$

and setting

$$\sigma_0 = \frac{S_0^2}{K} \tag{5.77}$$

one has

$$v = \sum_i \sum_j \frac{f_{ij} V_j \sigma_0}{\Delta_i + \sigma_0} \tag{5.78}$$

It appears from this equation that the non-Michaelian character of this expression originates from a spatial organization of the fixed charges. Let v_{ij} be the rate equation pertaining to cluster i, one has

$$v_{ij} = \left(<V> + \varepsilon_j \right) \frac{\sigma_0}{<\Delta> + \delta_i + \sigma_0} \tag{5.79}$$

where $<V>$ is the mean of V_j. This equation may be expanded in Taylor series with respect to the variable δ_i. One has then

$$v_{ij} = \left\{ <V> + \varepsilon_j \right\} \left\{ \frac{\sigma_0}{<\Delta> + \sigma_0} - \frac{\delta_i \sigma_0}{\left(<\Delta + \sigma_0 \right)^2} + \frac{\delta_i^2 \sigma_0}{\left(<\Delta> + \sigma_o \right)^3} - \ldots\ldots \right\} \tag{5.80}$$

and defining the dimensionless variables $\sigma_0^*, \delta_j^*, \varepsilon_j^*$ as

$$\sigma_0^* = \frac{\sigma_0}{<\Delta>}$$

$$\delta_j^* = \frac{\delta_j}{<\Delta>} \tag{5.81}$$

$$\varepsilon_j^* = \frac{\varepsilon_j}{<\Delta>}$$

equation (5.80) can be simplified to

$$v_{ij} = <V> \left(1 + \varepsilon_j^*\right) \left\{ \frac{\sigma_0^*}{1 + \sigma_0^*} - \frac{\delta_i^* \sigma_0^*}{\left(1 + \sigma_0^*\right)^2} + \frac{\delta_i^{*2} \sigma_0^*}{\left(1 + \sigma_0^*\right)^3} - \ldots\ldots \right\} \qquad (5.82)$$

The overall reaction rate may be obtained after summing up the rates v_{ij}. One finds

$$v = \sum_i \sum_j <V> f_{ij} \left(1 + \varepsilon_j^*\right) \left\{ \frac{\sigma_0^*}{1 + \sigma_0^*} - \frac{\delta_i^* \sigma_0^*}{\left(1 + \sigma_0^*\right)^2} + \frac{\delta_i^{*2} \sigma_0^*}{\left(1 + \sigma_0^*\right)^3} - \ldots\ldots \right\} \qquad (5.83)$$

As δ_i^* is small this series is of necessity convergent. Moreover one can write from this expression

$$\frac{v}{N(v)} = \frac{\sigma_0^*}{1 + \sigma_0^*} \left\{ 1 - \frac{\delta_i^*}{\left(1 + \sigma_0^*\right)} + \frac{\delta_i^{*2}}{\left(1 + \sigma_0^*\right)^2} - \ldots\ldots \right\} \qquad (5.84)$$

If the Taylor series converges rapidly the above expression (5.84) reduces to

$$\frac{v}{N <V>} = \frac{\sigma_0^*}{1 + \sigma_0^*} + \Xi(\sigma_0^*) \qquad (5.85)$$

where the function $\Xi(\sigma_0^*)$ is

$$\Xi(\sigma_0^*) = -\frac{\sigma_0^*}{\left(1 + \sigma_0^*\right)^2} \mathrm{cov}(\delta^*, \varepsilon^*) + \frac{\sigma_0^*}{\left(1 + \sigma_0^*\right)^3} \mathrm{var}(\delta^*) \qquad (5.86)$$

If $\Xi(\sigma_0^*)$ is plotted *versus* σ_0^* it may show first an enhancement followed by a decrease of the reaction rate. These effects are generated by the spatial organization of the enzyme relative to the charges (Fig. **7**). Moreover if the charges are homogeneously distributed in the matrix whereas the enzyme molecules are clustered the term in $\mathrm{cov}(\delta^*, \varepsilon^*)$ vanishes and equation (5.85) reduces to

$$\frac{v}{N<V>} = \frac{\sigma_0^*}{1+\sigma_0^*} + \frac{\sigma_0^*}{(1+\sigma_0^*)^3}\,\mathrm{var}(\delta^*) \tag{5.87}$$

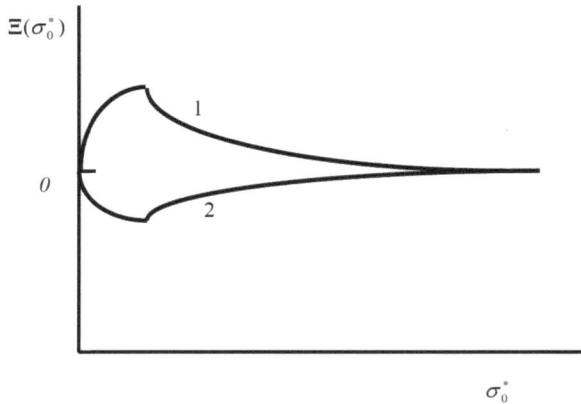

Figure 7: Two possible effects generated by the spatial organization of fixed charges relative to the enzyme molecules. The function $\Xi(\sigma_0^*)$ expresses how the spatial organization of the cell wall may enhance, or decrease, the reaction rate v. High values of var(δ^*) tend to increase $\Xi(\sigma_0^*)$ (curve 1) and high values of cov(δ^*, ε^*) produce the opposite effect.

3.5- Enzymes and Plant Cell Wall Extension

The problem of plant cell wall extension is a good illustration of the part played by the complexity of the interactions between enzymes and their charged environment. As a matter of fact, the above reasoning can be applied to a complex system which is involved in plant cell wall extension. Rigid envelopes of primary plant cell walls are made up of interconnected cellulose microfibrils. Cellulose is a linear glucan made up of glucopyranose units linked by $\beta(1 \to 4)$ bonds. Several glucan chains can be associated through hydrogen bonds to form cellulose microfibrils. Moreover many enzymes are ionically bound to plant cell walls, some of them being involved in plant cell wall extension. Extension and building up of the cell wall involve both the sliding of cellulose microfibrils and the incorporation into the wall of new polysaccharide material.

During cell wall extension, the fixed charge density of the wall is maintained which implies that the cell wall pectins undergo partial demethylation thank to a pectin methylesterase [30-33]. Hence this enzyme is responsible for the building up of the $\Delta\Psi$ of the rigid envelopes made up of interconnected cellulose microfibrils. Cellulose is a linear glucan made up of glucopyranose units linked by $\beta(1 \to 4)$ bonds. Several glucan chains may be associated through hydrogen bonds as to form cellulose microfibrils. One can schematically describe the extension of a micro-domain of the cell wall in the following way. One can tentatively assume that micro-domains of the cell wall occur under two states: a state called X_1 where most of acid groups of the pectins are ionized and a state, X_2M, where most of these groups are methylated. The conversion $X_1 \to X_2M$ requires the creep of cellulose microfibrils leading in turn to cell wall extension. This process is under the control of another enzyme, an endotransglucosylase, and is accompanied by the incorporation, in the extending region of the wall, of methylated proteins. As could be expected, the reverse process $X_2M \to X_1$ requires the pectin methyl esterase. The overall process may be described by a monocyclic cascade, as shown in Fig. **8**.

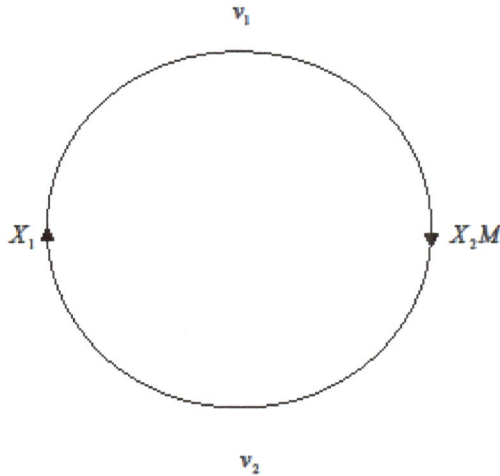

Figure 8: The two-state model of cell wall extension. The $X_1 \to X_2M$ is associated with cell wall extension and the reverse process, $X_2M \to X_1$, controlled by pectin methyl esterase generates the $\Delta\Psi$ values of the cell wall (see text).

Since high endotransgucosylase activity is generated by carboxylate groups in the environment of the enzyme, one must conclude that the optimum pH should be acidic. Alternatively, high pectin methylesterase activity takes place when pectins are methylated. One should conclude that the optimum pH of this enzyme should be close to neutrality. One can conclude that an extending region of the wall possesses many micro-domains with pectins in both ionized and methylated states. If one considers the $X_1 \rightarrow X_2 M$ process, it requires both the existence of X and of the endotransglucosylase. During the corresponding extension process, neutral compounds are incorporated in X_1 that is being converted in $X_2 M$. The process $X_2 M \rightarrow X_1$ describes the demethylation of $X_2 M$ leading in turn to the regeneration of X_1. If, in an expanding region of the wall, the system is in steady state the total density of micro-domains, $d(X)$, should remain constant and one has [36]

$$d(X) = d(X_1) + d(X_2 M) \qquad (5.88)$$

This is precisely the situation shown in Fig. **8**. Equation (5.88) can be rewritten as

$$\frac{d(X_2 M)}{dX} = 1 - \frac{d(X_1)}{d(X)} \qquad (5.89)$$

If we consider X_1 for instance, it carries a number of fixed charges. The corresponding charge density is δ_1. Among these charges, some of them are identical, for instance the carboxylate groups with a corresponding charge z' that can be considered some kind of valence. One has

$$d(X_1) = \frac{\delta_1}{z_1'} \qquad (5.90)$$

$$d(X_2 M) = \frac{\delta_2}{z_2'}$$

It then follows that

$$d(X) = \frac{\delta_1}{z_1'} + \frac{\delta_2}{z_2'} = \frac{z_2' \delta_1 + z_1' \delta_2}{z_1' z_2'} \qquad (5.91)$$

$$\frac{d(X_2 M)}{d(X)} = \frac{z_1' \delta_2}{z_2' \delta_1 + z_1' \delta_2} = \frac{\delta_2}{\dfrac{z_2'}{z_1'} \delta_1 + \delta_2} = \delta_2^* \qquad (5.92)$$

and one should have in the same way

$$\frac{d(X_1)}{d(X)} = \frac{\delta_1 z_2'}{z_2' \delta_1 + z_1' \delta_2} = \frac{\delta_1}{\delta_1 + \frac{z_1'}{z_2'} \delta_2} = \delta_1^* \tag{5.93}$$

If now we assume, for simplicity, that the system follows hyperbolic kinetics one has

$$v_1 = \frac{\tilde{V}_1 d(X_1)}{K_1 + d(X_1)} \tag{5.94}$$

$$v_2 = \frac{\tilde{V}_2 d(X_2 M)}{K_2 + d(X_2 M)}$$

where $\tilde{V}_1$ and $\tilde{V}_2$ are the apparent maximum rates, K_1 and K_2 the corresponding apparent Michaelis constants.

In this model, wall loosening enzymes, such as the endotransglycosylases, tend to generate a decrease of the charge density for they generate both loosening and extension of the wall together with the incorporation of neutral precursors. Alternatively, pectin methyl esterase must be involved in the building up of the electrostatic potential of the wall. In order to understand how pectin methyl esterase, in this system, is controlled by the local pH one has to express first how equations (5.94) depend upon δ_1^* *i.e.* the charge density generated by the enzyme. One has then

$$v_1 = \frac{\tilde{V}_1 \delta_1^*}{K_1^* + \delta_1^*} \tag{5.95}$$

and

$$v_2 = \frac{\tilde{V}_2 (1 - \delta_1^*)}{K_2^* + (1 - \delta_1^*)} \tag{5.96}$$

where K_1^* and K_2^* are defined as $K_1^* = K_1 / d(X)$ and $K_2^* = K_2 / d(X)$. As the overall system is assumed to be in steady state one has

$$\frac{\tilde{V}_1 \delta_1^*}{K_1^* + \delta_1^*} = \frac{\tilde{V}_2 (1 - \delta_1^*)}{K_2^* + (1 - \delta_1^*)} \tag{5.97}$$

If, as postulated in equations (5.95)-(5.97), $\tilde{V}_1$ and $\tilde{V}_2$ are pH-dependant but not K_1^* and K_2^* equation (5.97) can be rewritten as

$$\left(\frac{\tilde{V}_2}{\tilde{V}_1} - 1 \right) \delta_1^{*2} - \left\{ \left(\frac{\tilde{V}_2}{\tilde{V}_1} - 1 \right) - \tilde{V}_2 \left(\frac{K_1^*}{\tilde{V}_1} + \frac{K_2^*}{\tilde{V}_2} \right) \right\} \delta_1^* - \frac{K_1^*}{\tilde{V}_1} \tilde{V}_2 = 0 \tag{5.98}$$

This equation is similar to equation (5.35) and can display an all-or-none type of response when plotting δ_1^* as a function of pH_i. The existence of an abrupt transition of charge density is the consequence of opposite pH-dependence of pectin methyl esterase and of wall-loosening enzymes. Hence it is remarkable that this kind of behaviour is not the property of a specific enzyme but of a monocyclic cascade. In this overall cascade process pectin methyl esterase should be controlled by cations and protons and should be responsible for the building up of the difference of potential between the inside and outside of the cell wall. Moreover there should exist an ionic control of wall loosening enzymes.

4- GENERAL CONCLUSIONS

Classical biochemistry and molecular biology have explicitly, or implicitly, accepted the view that the individual knowledge of proteins and enzymes was, to a large extent, sufficient to understand the dynamics of metabolic processes. Modern studies have shown this is far from being the case. There are several reasons for such a situation. In the living cell enzyme reactions are not isolated processes. They constitute networks that possess an elasticity and, more generally, novel properties that have little to do with the individual behaviour of isolated enzyme reactions. Such networks possess their own global properties. Regulation of complex metabolic processes is not the property of a definite enzyme but rather some kind of "molecular democracy" that involves all the enzymes of this metabolic process.

A surprising response of apparently simple biochemical systems which cannot be observed with isolated enzymes is a all-or-none type of response to signals. Such a system requires the existence of, at least, two proteins that undergo reversible conversion thanks to two enzyme reactions. Processes of this type are apparently common during the process of plant cell wall extension. Numerous co-operative responses take place during metabolic processes whereas co-operativity is not observed with isolated enzymes. It may be due, for instance, to electrostatic attraction or repulsion of charged metabolites by complex cellular edifices such as plant cell wall. All these effects briefly described in the present Chapter have little to do with classical enzymology and still possess a major importance in the process of understanding biological phenomena in physical terms.

REFERENCES

[1]   Kacser, H. and Burns, J.A. (1979) Molecular democracy. Who shares the control? Biochem. Soc. Trans. 7, 1149-1160.

[2]   Kacser, H. and Burns, J. A. (1972) The control of flux. Symp. Soc. Exp. Biol. 27, 65-104.

[3]   Kacser, H. (1983) The control of enzyme systems *in vivo*. Elasticity analysis of the steady state. Biochem. Soc. Trans. 11, 35-40.

[4]   Heinrich, R. and Rapoport, T. A. (1974) A linear steady state treatment of enzymatic chains. Critique of the crossover theorem and a general procedure to identify interaction sites with an effector. Eur. J. Biochem. 42, 97-105.

[5]   Heinrich, R., Rapoport, S.M. and Rapoport, T.A. (1977) Metabolic regulation and mathematical models. Progress Biophys. Mol. Biol. 32, 1-82.

[6]   Giersch, C. (1988a) Control analysis of metabolic networks. 1 Homogeneous functions and the summation theorems for control coefficients. Eur. J. Biochem. 174, 509-513.

[7]   Giersch, C. (1988b) Control analysis of metabolic networks. 2 Total differentials and general formulation of the connectivity relations. Eur. J. Biochem. 174, 515-519.

[8]   Westeroff, H.V. and Chen, Y.D. (1984) How do enzyme activities control metabolite concentrations? Eur. J. Biochem. 142, 425-430.

[9]   Fell, D.A. and Sauro, H.M. (1985) Metabolic control and its analysis. Additional relationships between elasticities and control coefficients. Eur. J. Biochem.148, 555-561.

[10] Savageau, M.A. (1969) Biochemical systems analysis. I. Some mathematical properties of the rate laws for the component enzymatic reaction. J. Theor. Biol. 25, 365-369.

[11] Savageau, M.A. (1972) The behavior of intact biochemical control systems. Curr. Top. Cell. Regul. 6, 63-130.

[12] Chock, P.B., Rhee, S.G. and Stadtman, E. R. (1980a) Covalently interconvertible cascade systems. Methods Enzymol. Vol. 64, Part B Academic Press, New York, pp.297-325

[13] Chock, P.B., Rhee, S.G. and Stadtman, E. R. (1980b) Interconvertible enzyme cascades in cellular regulation. Annu. Rev. Biochem. 49, 813-843.

[14] Chock, P.B. and Stadtman, E.R. (1977) Superiority of interconvertible enzyme cascades in metabolic regulation: analysis of multicyclic systems. Proc. Natl. Acad. Sci. USA 74, 2766-2770.

[15] Rhee, S.G, Park, R., Chock, P.B. and Stadtman, E.R. (1978) Allosteric regulation of monocyclic interconvertible enzyme cascade systems: use of glutamine synthetase as an experimental model. Proc. Natl. Acad. Sci. USA, 75, 3138-3142.

[16]   Mura, U., Chock, P.B. and Stadtman, E. R. (1981) Allosteric regulation of the state of adenylation of glutamine synthetase in permeabilized cell preparations of *Escherichia coli.* J. Biol. Chem. 256, 13022-13029.

[17]   Rhee, S.G., Chock, P. B. and Stadtman, E. R. (1989) Regulation of *Escherichia coli* glutamine synthetase. Adv. Enzymol. 62, 37-92.

[18]   Stadtman, E. R. and Chock, P. B. (1977) Superiority of interconvertible enzyme cascades in metabolic regulation analysis of monocyclic cascades. Proc. Acad. Sci. USA 74, 2761-2766.

[19]   Goldbeter, A. and Koshland, D. E. (1981) An amplified sensitivity arising from covalent modification in biological systems. Proc. Natl. Acad Sci. USA 78, 6840-6844.

[20]   Goldbeter, A. and Koshland, D. E. (1984) Ultrasensitivity in biochemical systems controlled by covalent modifications. J. Biol. Chem.259, 14441-14447.

[21]   Goldbeter, A. and Koshland, D. E. (1982) Sensitivity amplification in biochemical systems. Quarterly Rev. Biophys.15, 555-591.

[22]   Ricard, J. (1989) Modulation of enzyme catalysis in organized biological systems. A physico-chemical approach. Catalysis Today 5, 275-384.

[23]   Ricard, J., Mulliert, G., Kellershohn, N. and Giudici-Orticoni, M. T. (1994) Dynamics of enzyme reactions and metabolic networks in living cells. A physico-chemical approach. Progress Mol. Subcell. Biol. 13,1-80.

[24]   Maurel, P. and Douzou,P. (1976) Catalytic implications of electrostatic potentials: the lytic activity of lysozyme as a model. J. Mol. Biol. 102, 253-264.

[25]   Ricard, J., Noat, G., Crasnier, M. and Job. D. (1981) Ionic control of acid phosphatase bound to plant cell walls. Plant Cell Environ. 3, 225-229.

[26]   Ricard, J., Kellershohn, N. and Mulliert, G. (1989) Spatial order as a source of kinetic cooperativity in organized bound enzyme systems. Biophys. J. 56, 477-487.

[27]   Crasnier,M., Moustacas, A.M. and Ricard, J. (1985)Electrostatic effects and calcium ion concentration as modulators of acid phosphatase bound to plant cell walls. Eur. J. Biochem. 151, 187-190.

[28]   Crasnier, M., Noat, G. and Ricard, J. (1980) Purification and molecular properties of acid phosphatase from sycamore cell walls. Plant Cell Environ. 3, 217-224.

[29]   Ricard, J., Kellershohn, N. and Mulliert, G. (1992) Dynamic aspects of long distance functional interactions between membrane bound enzymes. J. Theor. Biol. 156, 1-40.

[30]   Moustacas, A. M., Nari, J., Noat, G., Crasnier, M., Borel, M. and Ricard, J. (1986) Electrostatic effects and the dynamics of enzyme reactions at the surface of plant cells. 2. The role of pectin methyl esterase in the modulation of electrostatic effects in soyabean cell walls. Eur. J. Biochem. 155, 191-197.

[31]   Nari, J., Noat, G., Diamantidis; G., Woudstra, M. and Ricard, J. (1986) Electrostatic effects and the dynamics of enzyme reactions at the surface of plant cell. 3. Interplay between limited cell autolysis pectin methyl esterase activity and electrostatic effects in soybean cell walls. Eur. J. Biochem. 155, 199-202.

[32]   Nari, J., Noat, G. and Ricard, J. (1991) Pectin methyl esterase, metal ions and plant cell wall extension. Hydrolysis of pectin by plant cell wall pectin methyl esterase. Biochem. J. 279,342-350

[33]   Moustacas, A.M., Nari, J., Borel, M., Noat, G. and Ricard, J. (1991) Pectin methyl esterase, metal ions and plant cell wall extension. Biochem. J. 279, 351-354.

[34]   Ricard, J. and Noat, G. (1984) Enzyme reactions at the surface of living cells. I. Electric repulsion effects of charged ligands and recognition of signals from the external milieu. J. Theor; Biol. 109, 555-568.

[35]   Ricard, J. and Noat, G. (1984) Enzyme reactions at the surface of plant cells. II Destabilization in the membrane and conduction of signals. J. Theor. Biol. 109, 571-580.

[36]   Ricard, J. (1999) Biological complexity and the dynamics of life processes. Elsevier, Amsterdam, Lausanne, New York.

Send Orders for Reprints to reprints@benthamscience.net
Biological Systems: Complexity and Artificial Life, 2014, 99-125 99

CHAPTER 6

Thermodynamics of Energy Conversion within the Cell

Abstract: Many enzyme reactions take place in the living cell. Most of them are disfavored by thermodynamics *i.e.* most of them could not take place alone, in isolation. In the same vein, most of the processes involving the transport of matter are thermodynamically disfavored. It is only because these chemical and transport processes are coupled as to form a *system* that this system is feasible and functioning.

Keywords: Disfavored chemical reactions and cell compartments, Energy coupling, Active and passive transport, Transport processes, Fractionation factors, Fluxes in living cells, Scalar processes, Vectorial processes, Coupling between scalar and vectorial processes.

Diffusion and chemical reactions follow the rules of thermodynamics. At first sight, however, it seems that chemical reactions in free solution are different from those taking place in a living cell. There are in fact different reasons that explain why an enzyme usually does not behave the same way in dilute solution and *in vivo*, bound to cell organelles. Many enzymes, as well as small molecules and ions, are compartmentalized within the living cell. A second reason for the difference of behavior between a bound and a soluble enzyme relies upon diffusional resistances within the living cell. In free solution, diffusion of substrates and products to and from an enzyme's active site is usually an extremely rapid process relative to the steps of enzyme reaction. In fact, it may appear unthinkable that ions and molecules can accumulate in a living cell "uphill" an electrochemical gradient. It seems unconceivable that a chemical reaction involving a free energy increase can routinely take place in a living cell. As we shall see in this Chapter, these fundamental processes routinely take place in the living cell without violating any thermodynamic principle [1-22].

1- OCCURRENCE OF DISFAVOURED REACTIONS AND CELL COMPARTMENTS

Let us consider the chemical reaction

$$\nu_1 X_1 + \dots + \nu_f X_f + \dots \nu_k X_k \leftrightarrow \nu_{k+1} X_{k+1} + \dots + \nu_m X_m + \dots + \nu_p X_p$$

where the v's are the stoichiometric coefficients of reactants and products, respectively. If n_r and n_p are the numbers of reactants and products and ξ the advancement of the reaction one has

$$-dn_f = v_f d\xi \tag{6.1}$$

$$dn_p = v_p d\xi$$

If the reaction takes place in a close system at constant volume one has

$$dS = -\frac{1}{T}dG = \frac{1}{T}\left(\sum_f \mu_f dn_f + \sum_m \mu_m dn_m\right) \tag{6.2}$$

where S, G, T and μ are the entropy, the Gibbs free energy, the absolute temperature and the chemical potentials, respectively. Equation (6.2) can be rewritten as

$$dS = -\frac{1}{T}dG = \frac{1}{T}d\xi\left(\sum_f v_f \mu_f - \sum_m v_m \mu_m\right) \tag{6.3}$$

Moreover the affinity of the reaction is

$$A = \sum_f v_f \mu_f - \sum_m v_m \mu_m = \left(\frac{\partial G}{\partial \xi}\right)_{T,V} \tag{6.4}$$

and it follows that the rate of entropy production is

$$T\frac{dS}{dt} = A\frac{d\xi}{dt} \tag{6.5}$$

As the rate of entropy production cannot be negative, it follows that

$$A\frac{d\xi}{dt} \geq 0 \tag{6.6}$$

This expression is the so-called De Donder inequality. It states that during the spontaneous conversion of reactants to products in a closed system the affinity is of necessity positive.

Let us now consider a permeable membrane that separates the available space in two compartments, called *cis* (') and *trans* (''), of equal volume but having different absolute temperatures, T' and T''. We assume now that each of these compartments contains different chemical species having mole numbers n_i' and n_i'' and chemical potentials μ_i' and μ_i'' (Fig. **1**). Such systems display transport of matter and energy through the membrane. In the figure below it is assumed that the two transport processes takes place in the same direction which is indeed not compulsory.

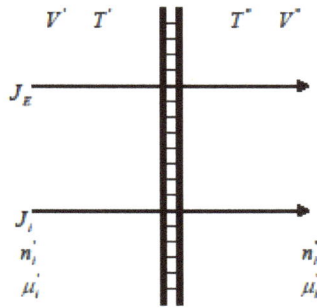

Figure 1: Transfer of matter and energy through a membrane. J_E and J_i are the fluxes of energy and matter through the membrane (see text).

The classical Gibbs-Duhem equations for the two compartments are

$$dE' = T'dS' + \sum_i \mu_i' dn_i' \tag{6.7}$$

$$dE'' = T''dS'' + \sum_i \mu_i'' dn_i''$$

Here, E' and E'' are the internal energies in the *cis* and *trans* compartments, S' and S'' the corresponding entropies in these compartments. As the entropy of the system is an extensive function

$$dS = dS' + dS'' \tag{6.8}$$

In this expression, dS is the entropy of the whole system. Conservation of matter and energy requires that

$$dn_i' = -dn_i''$$ (6.9)

$$dE' = -dE''$$

It then follows that equations (6.7) assume the form

$$dS' = -\frac{1}{T'}dE'' + \frac{1}{T'}\sum_i \mu_i' dn_i''$$ (6.10)

$$dS'' = \frac{1}{T''}dE'' - \frac{1}{T''}\sum_i \mu_i'' dn_i''$$

Hence expression (6.8) above can be rewritten as

$$dS = dS' + dS'' = \left(\frac{1}{T''} - \frac{1}{T'}\right)dE'' + \sum_i \left(\frac{\mu_i'}{T'} - \frac{\mu_i''}{T''}\right)dn_i''$$ (6.11)

and the rate of entropy dissipation is then

$$\frac{dS}{dt} = \left(\frac{1}{T''} - \frac{1}{T'}\right)\frac{dE''}{dt} + \sum_i \left(\frac{\mu_i'}{T'} - \frac{\mu_i''}{T''}\right)\frac{dn_i''}{dt}$$ (6.12)

If now one defines flows of energy, J_E and matter, J_i

$$J_E = \frac{dE''}{dt}, \; J_i = \frac{dn''}{dt}$$ (6.13)

as well as forces, X_E and X_i that drive these flows

$$X_E = \frac{1}{T''} - \frac{1}{T'}$$ (6.14)

$$X_i = \frac{\mu_i'}{T'} - \frac{\mu_i''}{T''}$$

one can rewrite equation (6.12) as

$$\frac{dS}{dt} = J_E X_E + \sum_i J_i X_i \geq 0 \qquad (6.15)$$

This relationship is important for it shows that some events apparently incompatible with common sense are in fact perfectly possible. For instance, storage of energy in a cell compartment is possible provided it is accompanied with migration of matter from compartment to compartment. Alternatively, migration of molecules, or ions, "against" a concentration gradient is possible if it is accompanied by an expenditure of energy. As we shall see later, a number of events, which seem to violate both physical laws and common sense, are in fact perfectly possible provided they are compatible with relation (6.15). This fundamental relationship represents the thermodynamic basis for coupling spontaneous, or active, transport of matter with storage, or expenditure, of energy.

2- ENERGY COUPLING

On the basis of equation (6.15), one can distinguish three types of energy couplings called chemiosmotic coupling, osmochemical coupling and osmosmotic coupling. Chemiosmotic coupling is a coupling between an exergonic chemical reaction (or reactions) and an active transport of ions. An osmochemical coupling means that this sort of coupling takes place between a spontaneous process of ion transport and an endergonic disfavoured reaction. Last, in osmoosmotic coupling different ions may be exchanged through a membrane (Fig. **2**).

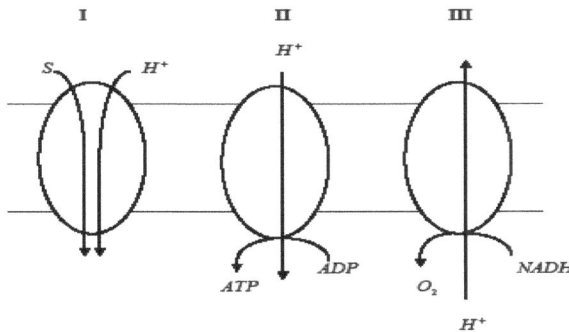

Figure 2: Different types of coupling processes. 1-Osmosmotic coupling is the spontaneous coupling between an exergonic chemical reaction and the active transport of ions (protons in the Figure). II-Osmochemical coupling is, for instance, the ATP synthesis coupled to proton transport. III-Chemiosmotic coupling is the transfer of an ion coupled to a chemical process.

3- PROCESSES OF ACTIVE AND PASSIVE TRANSPORT

As previously mentioned, transport processes can involve active or passive mechanisms. Let us assume that the available space in the cell can be separated in two compartments *cis* (') and *trans* ("). Let us, for instance, consider the migration of a ligand L from the *cis* to the *trans* compartment. If $[L']$ and $[L'']$ are the ligand concentrations in the two compartments, the corresponding electrochemical potentials, $\tilde{\mu}'_L$ and $\tilde{\mu}''_L$ are

$$\tilde{\mu}'_L = \mu^{\circ}_L + RT \ln[L'] + zF\psi' \tag{6.16}$$

$$\tilde{\mu}''_L = \mu^{\circ}_L + RT \ln[L''] + zF\psi''$$

where μ°_L is the standard chemical potential, ψ' and ψ'' the electrochemical potentials in the two compartments, F the Faraday constant and z the valence of the ion associated with a positive or a negative sign depending on the ion is a cation or an anion. The affinity of the diffusion process is defined as

$$A_L = -\Delta\tilde{\mu}_L = -(\tilde{\mu}''_L - \tilde{\mu}'_L) \tag{6.17}$$

if the transport takes place from *cis* to *trans* compartment. $\psi'' - \psi'$ is the electrostatic potential difference $\Delta\psi$. One has

$$\Delta\psi = \psi'' - \psi' \tag{6.18}$$

Under these conditions the affinity becomes

$$A_L = -\Delta\tilde{\mu}_L = -RT \ln\frac{[L'']}{[L']} - zF\Delta\psi \tag{6.19}$$

It follows that the condition required for spontaneous transport from *cis* to *trans* is

$$\frac{[L']}{[L'']} = \exp\left(\frac{A_L + zF\Delta\psi}{RT}\right) > 1 \tag{6.20}$$

which implies that

$$A_L + zF\Delta\psi > 0 \tag{6.21}$$

So far, the migration of L across the membrane was considered a passive event. This event, however, can be facilitated by the presence of carriers. Let us consider the situation described in Fig. **3** where a protein carrier, X, binds the ligand L and carry this ligand from the *cis* (') to the *trans* (") side of the membrane. The protein appears under two states, X' and X'', that may, or may not, bind ligand L.

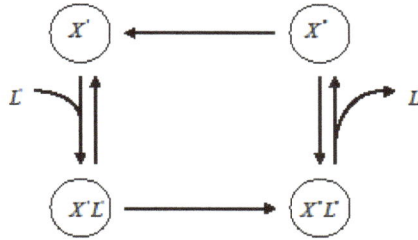

Figure 3: Active transfer of a ligand from the *cis* to the *trans* side of a membrane. See text.

In order to understand the physical mechanism of such a transport process one could represent this mechanism as shown in Fig. **4**. In this process, one has

$$Y' = X' + X'L = X'(1 + K_1'[L']) \tag{6.22}$$

$$Y'' = X'' + X''L = X''(1 + K_1''[L''])$$

where K_1' and K_1'' are the affinity constants of L for X' and X'', respectively. The dynamics of the conversion of Y' to Y'' and back (Fig. **4**) is controlled by functions f' and f'' defined by the following expressions

$$f_1' = \frac{X'}{X' + X'L} = \frac{1}{1 + K_1'[L']}$$

$$f_1'' = \frac{X''}{X'' + X''L} = \frac{1}{1 + K_1''[L'']} \tag{6.23}$$

$$f_2' = \frac{X'L}{X' + X'L} = \frac{K_1'[L']}{1 + K_1'[L']}$$

$$f_2'' = \frac{X''L''}{X'' + X''L''} = \frac{K_1''[L'']}{1 + K_1''[L'']}$$

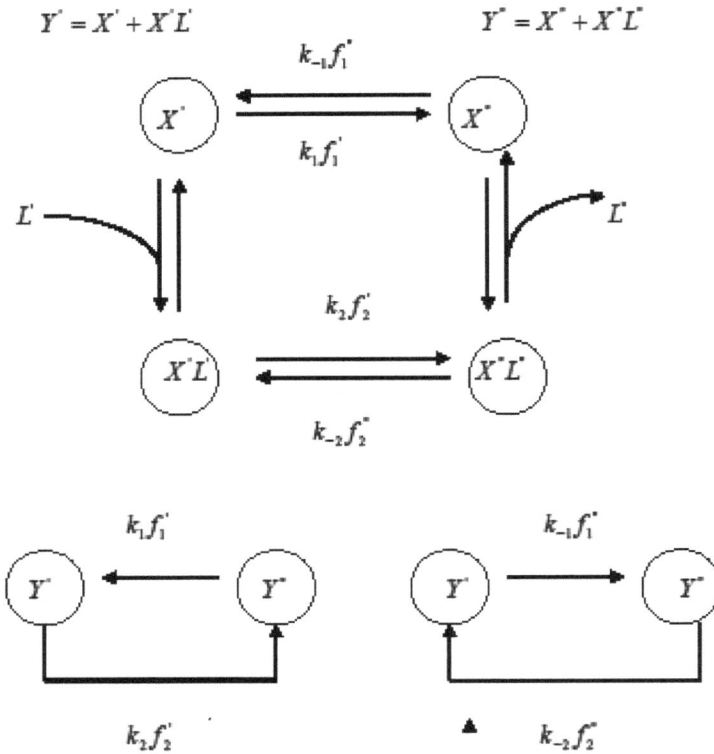

Figure 4: Definition of functions involved in a reversible transfer process. See text.

The kinetic scheme of Fig. **3** shows that ligand L is transferred from *cis* (') to *trans* (") if function G

$$G = \frac{k_1 f_1' k_{-2} f_2'' \left(\dfrac{k_2 f_2' k_{-1} f_1''}{k_1 f_1' k_{-2} f_2''} - 1 \right)}{k_{-1} f_1'' + k_2 f_2' + k_1 f_1' + k_{-2} f_2''} \tag{6.24}$$

adopts positive values. This function will be positive if

$$\frac{k_2 f_2' k_{-1} f_1''}{k_1 f_1' k_{-2} f_2''} > 1 \tag{6.25}$$

Moreover one has

$$\frac{k_2 f_2' k_{-1} f_1''}{k_1 f_1' k_{-2} f_2''} = \frac{K_2}{K_1} \frac{f_1''}{f_1'} \frac{f_2'}{f_2''} \tag{6.26}$$

and

$$\frac{f_1''}{f_1'} = \frac{1 + K_1'[L']}{1 + K_1''[L'']}$$

$$\frac{f_2'}{f_2''} = \frac{(1 + K_1''[L''])K_1'[L']}{(1 + K_1'[L'])K_1''[L'']} \tag{6.27}$$

$$\frac{f_1''}{f_1'} \frac{f_2'}{f_2''} = \frac{K_1'}{K_1''} \frac{[L']}{[L'']}$$

It follows that

$$\frac{k_2 f_2' k_{-1} f_1''}{k_1 f_1' k_{-2} f_2''} = \frac{K_2 K_1'}{K_1 K_1''} \frac{[L']}{[L'']} \tag{6.28}$$

Moreover thermodynamics requires that

$$K_2 K_1' = K_1 K_1'' \tag{6.29}$$

and one has

$$\frac{k_2 f_2' k_{-1} f_1''}{k_1 f_1' k_{-2} f_2''} = \frac{[L']}{[L'']} = \exp\left(-\frac{\Delta \tilde{\mu}_L - zF\Delta\Psi}{RT}\right) \tag{6.30}$$

4- TRANSPORT PROCESSES AND ENERGY COUPLING

The previous results imply that the biological membranes can couple transport processes to chemical reactions. The resulting events are called scalar-vectorial processes. Let us consider for instance a scalar-chemical reaction, for instance the hydrolysis of a substance S into P and Q

$$S + W \rightarrow P + Q$$

where W is water. This scalar reaction can drift the vectorial transport of S across a membrane as shown in Fig. **5**. A carrier X may be present on both sides of a membrane. The binding of S to a carrier X on the *cis* (') side of the membrane facilitates the vectorial transport of S to the *trans* (") side and its resulting hydrolysis into P and Q.

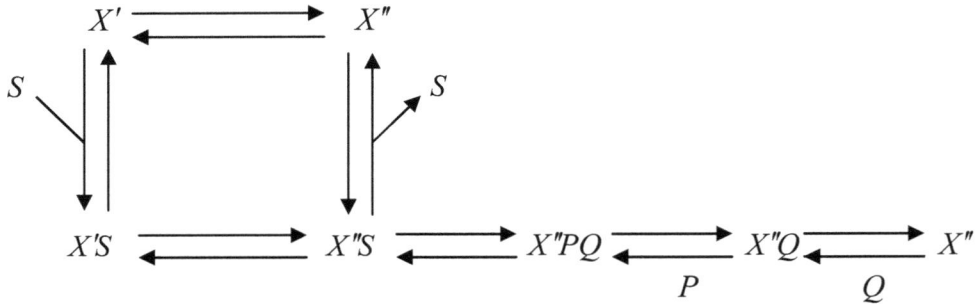

Figure 5: Energy release coupled to transport process. Transport of S through a membrane is coupled to its hydrolysis into P and Q.

The sequence of reactions that contribute to drive the transport of S across the membrane coupled to its conversion into P and Q is

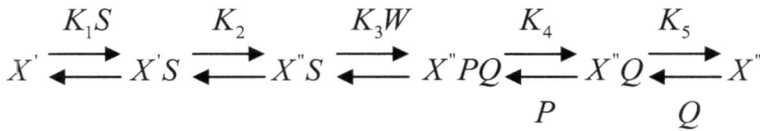

The apparent equilibrium constant of the coupled scalar-vectorial process is then

$$\tilde{K} = K_1 K_2 K_3 K_4 K_5 \frac{[S]}{[P][Q]} \tag{6.31}$$

It appears from Fig. **5** and from equation (6.31) that if $[P]$ and $[Q]$ are large the transport of S across the membrane is small.

The coupling between a transport and a chemical process taking place in a membrane is shown in Fig. **6**. The first type of event is the transport of a ligand facilitated by diffusion (Fig. **6A**). The second event is a transport of a ligand against a concentration gradient (Fig. **6B**) coupled to the hydrolysis of ATP. The third type of process is a transport of a ligand along an electrochemical gradient which allows endergonic synthesis of molecules such as ATP (Fig. **6C**). The last type of process is the occurrence of an endergonic chemical reaction coupled to a diffusion process (Fig. **6D**).

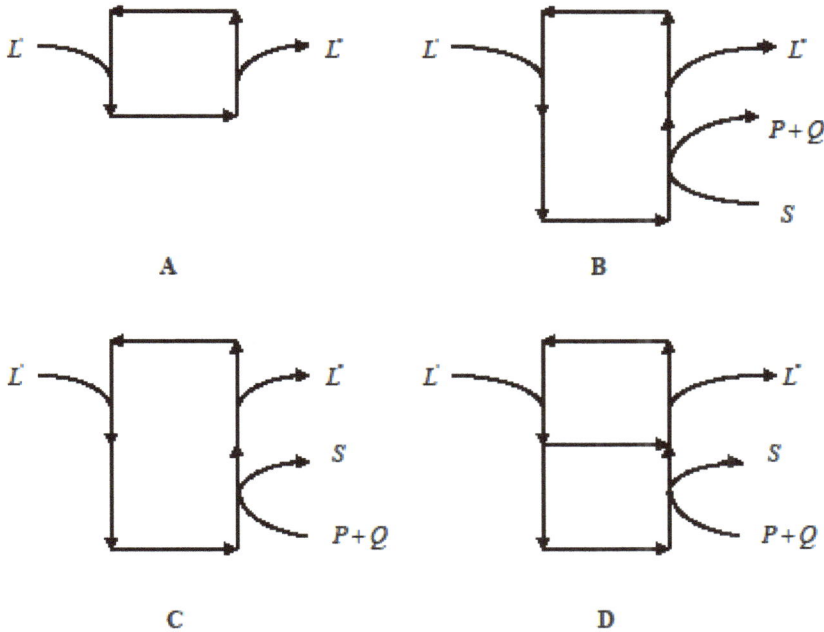

Figure 6: Coupling between different types of events. A. Facilitated diffusion of a ligand L through a membrane. B. Transport of a ligand against a concentration gradient thanks to the exergonic reaction $S \rightarrow P+Q$. C. and D. Endergonic reaction (for instance synthesis of ATP) coupled to the spontaneous transfer of L through the membrane. D.

A general kinetic model is shown in Fig. **7A** with its node contraction. In the process of node contraction we assume that parts of the system are in quasi-equilibrium (Fig. **7B**). This means that in the model of Fig. **7A** the nodes X', LX'_1 and LX'_2 on one hand, X'', LX''_1, LX''_2 on the other hand are in quasi-equilibrium and can be represented by two virtual species Y' and Y''. Hence one has

$$Y' = [X'] + [LX'_1] + [LX'_2] \qquad\qquad (6.32)$$

$$Y'' = [X''] + [LX''_1] + [LX''_2]$$

The general kinetic scheme of Fig. **7A** can then be schematized as shown in Fig. **7B**. The $f's$ that appear in this figure are fractionation factors. They express the respective contributions of the various enzyme states within Y' and Y'' upon assuming these states to be in pseudo-equilibrium. In the case of Fig. **7B** the expressions of the f' and f'' are defined as

$$f_1'' = \frac{[X'']}{[X'']+[X_1''L'']+[X_2''L'']} = \frac{1}{1+K_1''[L''](1+K_2'')}$$

$$f_2' = \frac{[X_1'L']}{[X']+[X_1'L']+[X_2'L_2']} = \frac{K_1'[L']}{1+K_1'[L'](1+K_2')} \tag{6.33}$$

$$f_3' = \frac{[X_2'L']}{[X']+[X_1'L']+[X_2'L']} = \frac{K_1'K_2'[L']}{1+K_1'[L'](1+K_2')}$$

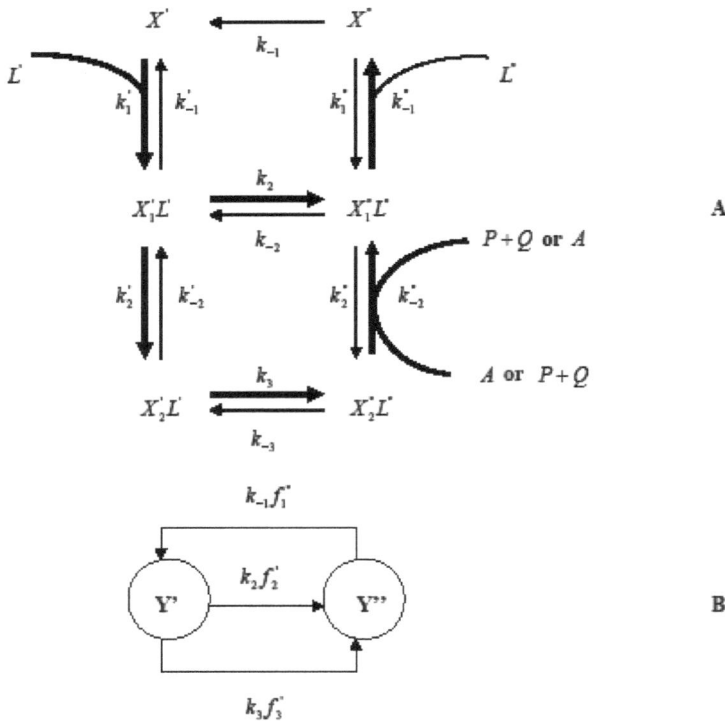

Figure 7: Condensation of a complex scheme into a simpler one. Condensation of scheme A (in the forward direction) into scheme B.

Only three $f's$ will appear in the corresponding equation because we are assuming that the step $X'' \to X'$ is nearly irreversible (Fig. **7**), which is likely to occur in order to explain that L is transferred, in an irreversible manner, from the *cis* (') to the *trans* (") side of the membrane. This is a situation which is likely to occur and that is often connected to the synthesis, or to the consumption, of ATP. The mutual expressions of f' and f'' are dependent upon the complexity of the model. Thus, for instance, if we consider the active, reversible, transport of a substance through a membrane without the occurrence of any other process, the kinetic scheme of Fig. **7** becomes simpler (Fig. **8**) and the expressions of f' and f'' are now.

$$f_1' = \frac{1}{1 + K_1'[L']}$$

$$f_2' = \frac{K_1'[L']}{1 + K_1'[L']} \tag{6.34}$$

$$f_1'' = \frac{1}{1 + K_1''[L'']}$$

$$f_2'' = \frac{K_1''[L'']}{1 + K_1''[L'']}$$

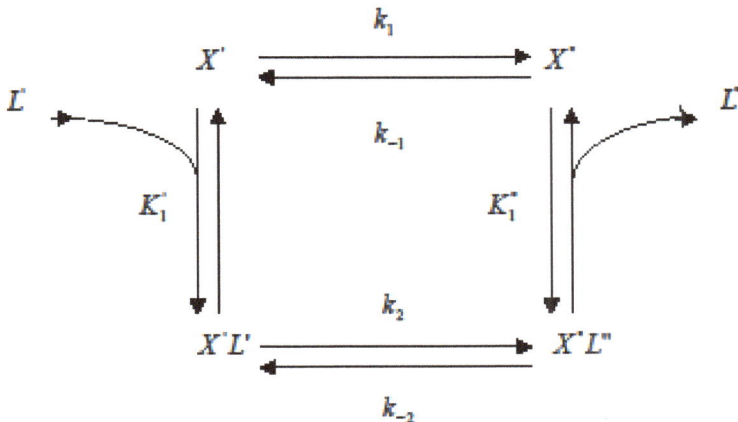

Figure 8: Active transport of L accross a membrane. See text.

If we consider for instance the reversible transport process of a substrate through a membrane the situation is depicted in Fig. **9**. The ligand L is bound to a receptor on the *cis*(') side of the membrane and transported to the *trans*(") side. The situation is depicted in Fig. **9**. We assume, as already pointed out, that the free receptor is in quasi-equilibrium with the receptor –ligand complex on the *cis*(') and *trans*(") sides of the membrane.

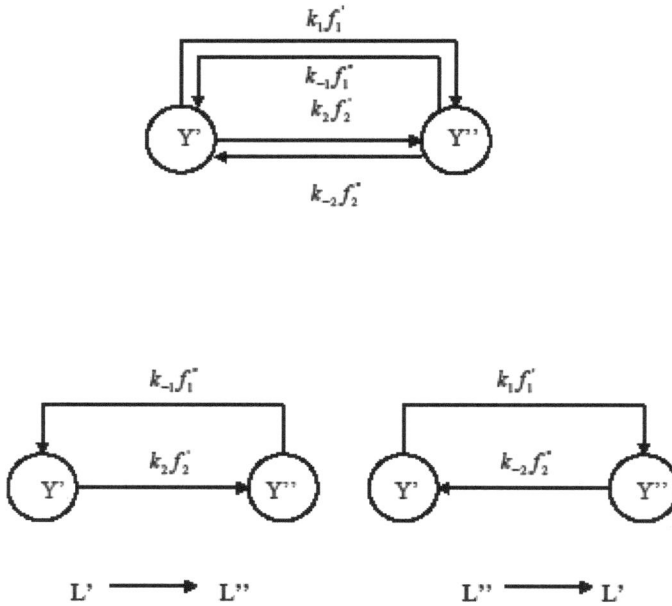

Figure 9: The forward and backward processes of the transport of L accross the membrane. See text.

The equation of flux is made up of a number of elementary events shown in Fig. **9**. The corresponding equation of the flux collects both rate constants in forward and backward directions as well as fractionation factors (equations 6.33 and 6.34). The flux equation is then

$$J = \frac{k_2 f_2' k_{-1} f_1'' - k_1 f_1' k_{-2} f_2''}{k_1 f_1' + k_{-2} f_2'' + k_2 f_2' + k_{-1} f_1''} \tag{6.35}$$

where the fractionation factors f' and f'' are defined in equations (6.34). Taking advantage of equations (6.34) the equation of the flux becomes

$$J = \frac{k_{-1}k_2 K_1'[L_1'] - k_1 k_{-2} K_1''[L_1'']}{(k_1 + k_2 + K_1'[L'] + k_{-1}K_1'[L'])(1 + K_1''[L'']) + k_{-2}K_1''[L''](1 + K_1'[L'])} \tag{6.36}$$

If the concentration of L on the *trans*(") side of the membrane is negligible, the equation of flux is that of a rectangular hyperbola (Fig. **10**).

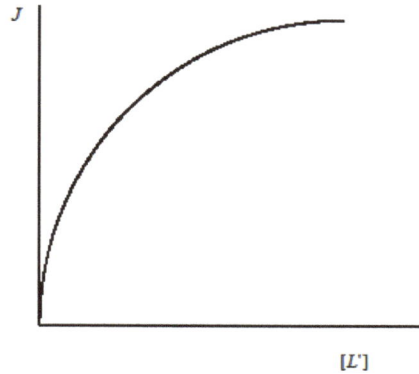

Figure 10: Typical curve of the transfer of L from the *cis* to the *trans* side of a membrane. In this scheme it is assumed that $\left[L' \right] \gg \left[L'' \right]$.

Let us now consider a realistic situation in which ATP (A in Figs. **7** and **10**) is being converted into ADP and phosphate (P and Q) and allows the irreversible transport of L across a membrane. One can also consider the symmetrical situation where it is the irreversible transport of L across the membrane that allows synthesis of ATP (A) from ADP(P) and phosphate (Q). These to situations are described by two equivalent models (Figs. **7** and **11**). In the first situation one is assuming that ATP hydrolysis is the motor of the active transport of L. In the second situation it is the spontaneous transport of L that allows the synthesis of ATP from ADP and phosphate.

If we assume that it is the conversion of ATP into ADP and phosphate that drives the active transport of L across the membrane the resulting equation is

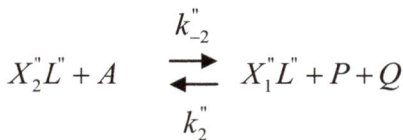

$$X_2''L'' + A \underset{k_2''}{\overset{k_{-2}''}{\rightleftarrows}} X_1''L'' + P + Q$$

Under these conditions the corresponding equilibrium is described by the following equation

$$[X_2^{"}L^{"}] = \frac{k_2^{"}}{k_{-2}^{"}} \frac{[P][Q]}{[A]} [X_1^{"}L^{"}]$$ (6.37)

and the apparent equilibrium constant $\tilde{K}_2^{"}$ is

$$\tilde{K}_2^{"} = K_2^{"} \frac{[P][Q]}{[A]}$$ (6.38)

If alternatively we assume that it is the spontaneous conversion of $X_2^{"}L^{"}$ into $X_1^{"}L^{"}$ that drives ATP synthesis from ADP and phosphate, one has

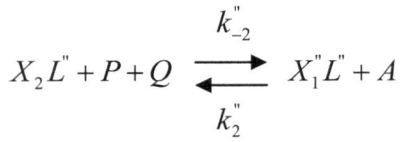

$$X_2 L^{"} + P + Q \underset{k_2^{"}}{\overset{k_{-2}^{"}}{\rightleftarrows}} X_1^{"}L^{"} + A$$

and the corresponding apparent equilibrium constant is then

$$\tilde{K}_2^{"} = K_2^{"} \frac{[A]}{[P][Q]}$$ (6.39)

In order to explain how active transport of L across a membrane can generate ATP hydrolysis, or alternatively, how ATP consumption can interfere with the active transport of L across a membrane, one can consider two sensible models such as the one already discussed (Fig. 7) and the one, slightly simpler, shown below (Fig. 11).

In these two models the f's are identical except that, in the present one, $f_2^{'}$ is lacking. Let us call A and B these two models. One can easily see that for model A the rate of transport of L across the membrane is

$$J = \frac{k_{-1} f_1^{"} (k_2 f_2^{'} + k_3 f_3^{'})}{k_{-1} f_1^{"} + k_3 f_3^{'} + k_2 f_2^{'}}$$ (6.40)

The terms of the denominator of this equation can be rewritten as

$$k_{-1}f_1'' = \frac{k_{-1}}{1+K_1''[L'](1+K_2'')} = \frac{k_{-1}\{1+K_1'[L'](1+K_2')\}}{\{1+K_1''[L'](1+K_2'')\}\{1+K_1'[L'](1+K_2')\}}$$

$$k_3f_3' = \frac{k_3K_1'K_2'[L']}{1+K_1'[L'](1+K_2')} = \frac{k_3K_1'K_2'[L']\{1+K_1''[L'](1+K_2'')\}}{\{1+K_1''[L'](1+K_2'')\}\{1+K_1'[L'](1+K_2')\}}$$

$$k_2f_2' = \frac{k_2K_1'[L']}{1+K_1'[L'](1+K_2')} = \frac{k_2K_1'[L']\{1+K_1''[L'](1+K_2'')\}}{\{1+K_1''[L'](1+K_2'')\}\{1+K_1'[L'](1+K_2')\}}$$

(6.41)

whereas the numerator assumes the form

$$k_{-1}f''(k_2f_2' + k_3f_3') = \frac{k_{-1}K_1'[L'](k_2 + k_3K_2')}{\{1+K_1''[L'](1+K_2'')\}\{1+K_1'[L'](1+K_2')\}}$$

(6.42)

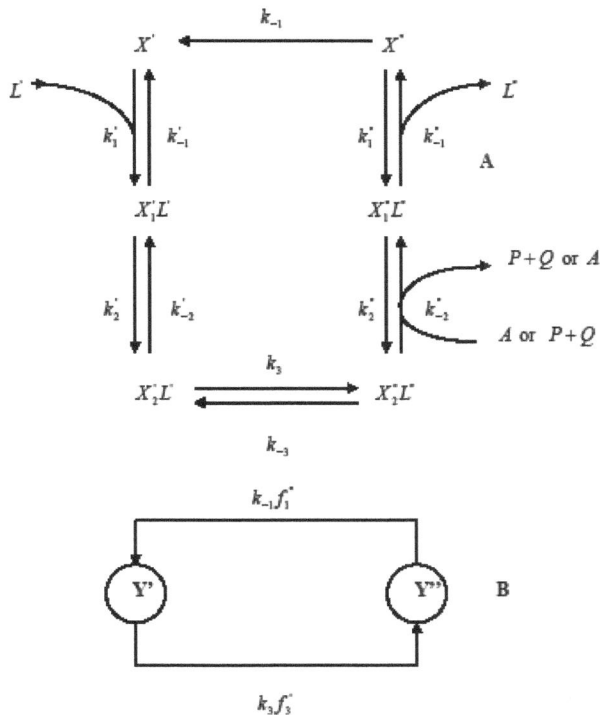

Figure 11: Transport of a ligand accross a membrane coupled to ATP synthesis, or hydrolysis. Top: The complete kinetic scheme. Bottom: The condensed kinetic scheme.

The expression of the flux J can be derived from the ratio of expression (6.42) and of the sum of expressions (6.41). One finds

$$J = \frac{k_{-1}\{1 + K_1'[L'](1 + K_2')\}}{k_{-1}\{1 + K_1'[L'](1 + K_2')\} + \{1 + K_1''[L''](1 + K_2'')\}\{k_2 + k_3 K_2'\}K_1'[L']} \quad (6.43)$$

It appears obvious from this equation that, when ATP is being converted into ADP and phosphate, the equilibrium constant K_2'' becomes small and the flux J increases. A similar conclusion is obtained if the conversion of $X_2'L'$ into $X_1''L''$ drives ATP synthesis. As shown in equation (6.39) the increase of ATP concentration produces a decrease of K_2'' thus producing an enhancement of flux J.

In the case of model B (Fig. **11**) the rate equation is simpler and the term in f_2' is lacking. One has

$$J = \frac{k_{-1} f_1'' k_3 f_3'}{k_{-1} f_1'' + k_3 f_3'} \quad (6.44)$$

with

$$k_{-1} f_1'' k_3 f_3' = \frac{k_{-1} k_3 K_1' K_2'[L']}{\{1 + K_1''[L''](1 + K_2'')\}\{1 + K_1'[L'](1 + K_2')\}} \quad (6.45)$$

and

$$k_{-1} f_1'' = \frac{k_{-1}\{1 + K_1'[L'](1 + K_2')\}}{\{1 + K_1''[L''](1 + K_2'')\}\{1 + K_1'[L'](1 + K_2')\}} \quad (6.46)$$

$$k_3 f_3' = \frac{k_3 K_1' K_2'[L']\{1 + K_1''[L''](1 + K_2'')\}}{\{1 + K_1''[L''](1 + K_2'')\}\{1 + K_1'[L'](1 + K_2')\}}$$

It follows that

$$J = \frac{k_{-1} k_3 K_1' K_2'[L']}{k_{-1}\{1 + K_1'[L'](1 + K_2')\} + k_3 K_1' K_2'[L']\{1 + K_1''[L''](1 + K_2'')\}} \quad (6.47)$$

which is to be compared to equation (6.43)

5- THERMODYNAMICS AND COUPLING OF SCALAR AND VECTORIAL PROCESSES

Let us consider the process considered in Fig. **12**. Some aspects of this process have already been discussed. They imply that a scalar chemical process, *viz.* the synthesis or the hydrolysis, of ATP is coupled to the vectorial transfer of a ligand from the *cis*(') to the *trans*(") side of a membrane. Depending on the process is oriented $L' \rightarrow L''$ or $L'' \rightarrow L'$ the flow within the membrane can be oriented in two different directions, as shown in Fig. **12**.

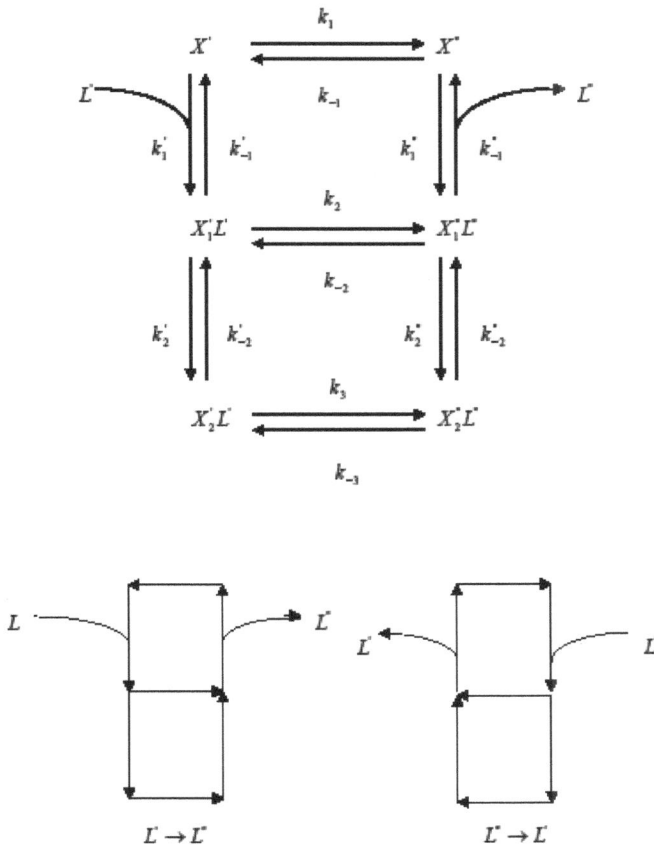

Figure 12: The same kinetic scheme can explain the transfer of a ligand in both directions. Top: The complete kinetic scheme. Bottom: The kinetic scheme take account of the two types of transfer $\Delta G_{C1}^{\neq'} = \Delta G_{C1}^{\neq*} - U_\tau' + U_\gamma'$ and $\Delta G_{C1}^{\neq*} - \Delta G_{-S1}^{\neq*}$.

If we assume, as done previously, that the states $X', X_1'L, X_2'L$ and $X'', X_1''L, X_2''L$ are in fast equilibrium, the model can be schematized by the two diagrams shown in Fig. **13**. The system A describes what is happening if $L' \to L''$ and the system B is a description of the process $L'' \to L'$.

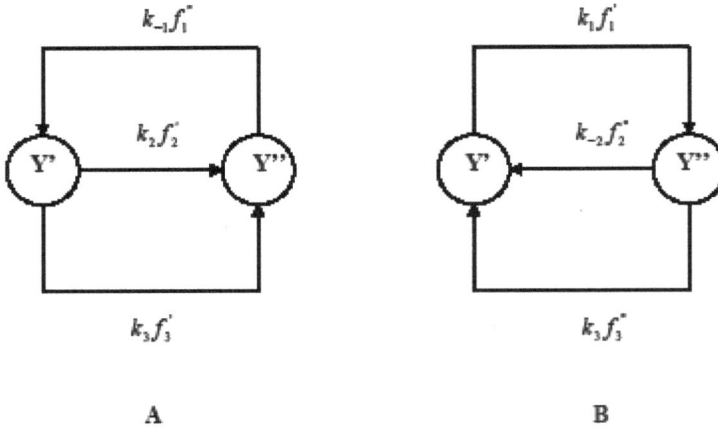

A   **B**

Figure 13: Condensation of the two kinetic schemes of Fig. **12**. A: Condensation of the kinetic scheme pertaining to $L' \to L''$. B: Condensation of the kinetic scheme pertaining to $L'' \to L'$.

As previously, one can define fractionation factors

$$f_1' = \frac{1}{1 + K_1'[L'](1 + K_2')} \quad f_1'' = \frac{1}{1 + K_1''[L''](1 + K_2'')}$$

$$f_2' = \frac{K_1'[L']}{1 + K_1'[L'](1 + K_2')} \quad f_2'' = \frac{K_1''[L'']}{1 + K_1''[L''](1 + K_2'')} \qquad (6.48)$$

$$f_3' = \frac{K_1' K_2'[L']}{1 + K_1'[L'](1 + K_2')} \quad f_3'' = \frac{K_1'' K_2''[L'']}{1 + K_1''[L''](1 + K_2'')}$$

In the direction $L' \to L''$ the flux J is defined as

$$J = \left\{ k_{-1} f_1'' (k_2 f_2' + k_3 f_3') - k_1 f_1' (k_{-2} f_2'' + k_{-3} f_3'') \right\} / \Delta \qquad (6.49)$$

where Δ collects the non-cyclic terms

$$\Delta = k_1 f_1' + k_2 f_2' + k_3 f_3' + k_{-1} f_1'' + k_{-2} f_2'' + k_{-3} f_3'' \tag{6.50}$$

Expression (6.49) can be rewritten as

$$J = \left\{ k_1 f_1' (k_{-2} f_2'' + k_{-3} f_3'') \right\} \left\{ \frac{k_{-1} f_1'' (k_2 f_2' + k_3 f_3')}{k_1 f_1' (k_{-2} f_2'' + k_{-3} f_3'')} - 1 \right\} / \Delta \tag{6.51}$$

Clearly the system will be in equilibrium if

$$\frac{k_{-1} f_1'' (k_2 f_2' + k_3 f_3')}{k_1 f_1' (k_{-2} f_2'' + k_{-3} f_3'')} = 1 \tag{6.52}$$

Let us consider the ratios $k_2 f_2' / k_{-2} f_2''$ and $k_3 f_3' / k_{-3} f_3''$, one has

$$\frac{k_2 f_2'}{k_{-2} f_2''} = K_2 \frac{K_1'[L']}{1 + K_1'[L'](1 + K_2')} \frac{1 + K_1''[L''](1 + K_2'')}{K_1''[L'']} \tag{6.53}$$

$$\frac{k_3 f_3'}{k_{-3} f_3''} = K_3 \frac{K_1' K_2'[L']}{1 + K_1'[L'](1 + K_2')} \frac{1 + K_1''[L''](1 + K_2'')}{K_1'' K_2''[L'']} \tag{6.54}$$

One can notice that the ratio $k_2 f_2' / k_{-2} f_2''$ will be equal to $k_3 f_3' / k_{-3} f_3''$ if

$$K_2 K_2'' = K_3 K_2' \tag{6.55}$$

Simple inspection of the model of Fig. **13** shows that this relationship is imposed by the first principle of thermodynamics. Hence one has

$$\frac{k_{-1} f_1'' (k_2 f_2' + k_3 f_3')}{k_1 f_1' (k_{-2} f_2'' + k_{-3} f_3'')} = \frac{k_{-1} f_1'' k_2 f_2'}{k_1 f_1' k_{-2} f_2''} = \frac{k_{-1} f_1'' k_3 f_3'}{k_1 f_1' k_{-3} f_3''} \tag{6.56}$$

Moreover one can derive simple expressions for the ratios $f_1'' / f_2'', f_2' / f_1', f_1'' / f_3''$ and f_3' / f_1'. One finds

$$\frac{f_1''}{f_2''} = \frac{1}{1 + K_1''[L''](1 + K_2'')} \frac{1 + K_1''[L''](1 + K_2'')}{K_1''[L'']} = \frac{1}{K_1''[L'']}$$

$$\frac{f_2'}{f_1'} = \frac{K_1'[L']\{1+K_1'[L'](1+K_2')\}}{1+K_1'[L'](1+K_2')} = K_1'[L']$$

$$\frac{f_1''}{f_3''} = \frac{1+K_1''[L''](1+K_2'')}{K_1''[L'']\{1+K_1''[L''](1+K_2'')\}} = \frac{1}{K_1''[L'']} \qquad (6.57)$$

$$\frac{f_3'}{f_1'} = \frac{K_1'K_2'[L']\{1+K_1'[L'](1+K_2')\}}{1+K_1'[L'](1+K_2')} = K_1'K_2'^2[L']$$

Taking advantage of these expressions one finds

$$\frac{k_{-1}f_1''k_2f_2'}{k_1f_1'k_{-2}f_2''} = \frac{K_2K_1'}{K_1K_1''}\frac{[L']}{[L'']} \qquad (6.58)$$

and

$$\frac{k_{-1}f_1''k_3f_3'}{k_1f_1'k_{-3}f_3''} = \frac{K_3K_2K_1'}{K_1K_1''}\frac{[L']}{[L'']} \qquad (6.59)$$

The first principle of thermodynamics requires that, in expression (6.57) above

$$K_2K_1' = K_1K_1'' \qquad (6.60)$$

It then follows that

$$\frac{k_{-1}f_1''k_2f_2'}{k_1f_1'k_{-2}f_2''} = \frac{[L']}{[L'']} \qquad (6.61)$$

Similarly, thermodynamics prescribes that

$$K_2'K_3 = K_2K_2'' \qquad (6.62)$$

and

$$K_1'K_2 = K_1K_2'' \qquad (6.63)$$

Hence expression (6.59) becomes

$$\frac{k_{-1}f_1''k_3f_3'}{k_1f_1'k_{-3}f_3''} = K_2''\frac{[L']}{[L'']} \tag{6.64}$$

As K_2'' is an apparent equilibrium constant that includes the concentrations of ATP (S), ADP (P), and phosphate (Q), one has

$$K_2'' = \frac{k_2''}{k_{-2}''}\frac{[S]}{[P][Q]} \tag{6.65}$$

or, depending on the situation

$$K_2'' = \frac{k_2''}{k_{-2}''}\frac{[P][Q]}{[S]} \tag{6.66}$$

and it becomes obvious that the vectorial transport process is linked to the scalar process of ATP synthesis or consumption.

As shown earlier in this Chapter, a transfer from *cis*(') to *trans*(") compartment is obtained if

$$\frac{[L']}{[L'']} = \exp\left\{-\frac{\Delta\tilde{\mu}_L - zF\Delta\psi}{RT}\right\} \tag{6.67}$$

Hence it follows that the active transport from the *cis*(') to the *trans* (") compartment requires that

$$-(\Delta\tilde{\mu}_L - zF\Delta\psi) > 0 \tag{6.68}$$

The flow in the forward direction is then

$$J^+ = J^-\left\{\exp\left(-\frac{\Delta\tilde{\mu}_L - zF\Delta\psi}{RT}\right)\right\} \tag{6.69}$$

where J^- is the flow in the backward direction. Let us assume for instance that ATP (S) is hydrolyzed into ADP (P) and phosphate (Q) the corresponding affinity of this reaction is

$$A_S = -\left(\Delta G_S^0 + RT \ln \frac{[P][Q]}{[S]} \right) \tag{6.70}$$

where ΔG_S^0 is the standard free energy of ATP hydrolysis. If the ratio $\rho = [P][Q]/[S]$ is chosen as to possess the value it had in the equilibrium constant K_2'' (equation 6.65) the affinity is, of necessity, positive. One has then

$$\frac{[P][Q]}{[S]} = \rho = \exp\left\{ -\frac{\Delta G_S^0 + A_S}{RT} \right\} \tag{6.71}$$

If it is the ratio $[S]/[P][Q]$ that appears in the expression of K_2'' (equation 6.65), the affinity is of necessity positive. One has then

$$\frac{[S]}{[P][Q]} = \frac{1}{\rho} = \exp\left\{ \frac{\Delta G_S^0 + A_S}{RT} \right\} \tag{6.72}$$

If relationship (6.70) applies, expression (6.64) can be rewritten as

$$\frac{k_{-1}f_1''k_3f_3'}{k_1f_1'k_{-3}f_3''} = \exp\left\{ \frac{\Delta G_S^0 + A_S - \left(\Delta \tilde{\mu}_L - zF\Delta\psi \right)}{RT} \right\} \tag{6.73}$$

and if it is relation (6.71), one has

$$\frac{k_{-1}f_1''k_3f_3'}{k_1f_1'k_{-3}f_3''} = \exp\left\{ \frac{-(\Delta G_S^0 + A_S) - (\Delta \tilde{\mu}_L - zF\Delta\psi)}{RT} \right\} \tag{6.74}$$

Active transport from *cis*(') to *trans*(") compartment requires that

$$\frac{k_{-1}f_1''k_3f_3'}{k_1f_1'k_{-3}f_3''} > 1 \tag{6.75}$$

and this implies that

$$\Delta G_S^0 + A_S - \left(\Delta \tilde{\mu}_L - zF\Delta\psi\right) > 0 \qquad (6.76)$$

for relation (6.71) or

$$-\left(\Delta G_S^0 + A_S\right) - \left(\Delta \tilde{\mu}_L - zF\Delta\psi\right) > 0 \qquad (6.77)$$

for relation (6.72). Of particular interest is the situation where the transport is an active process that takes place against an electro-chemical gradient. Under these conditions one has

$$zF\Delta\psi - \Delta\tilde{\mu}_L < 0 \qquad (6.78)$$

and active transport requires that

$$\Delta G_S^0 + A_S > 0 \qquad (6.79)$$

6- GENERAL CONCLUSIONS

The observation of living cells provides surprising results. One can observe, for instance, that a number of chemical reactions, which are considered as physically impossible in isolation, currently take place in the living cell. For decades, this kind of conclusion has led biologists to claim that life sciences cannot be reduced to physics and chemistry. In this perspective biological processes should be viewed in a vitalist perspective. It is evident today that such a conclusion should be considered hasty, or even invalid. There is little doubt that the approach of biological problems in terms of conventional biological chemistry and classical molecular biology takes into account *isolated* processes and, for that reason, leads to incorrect conclusions.

It appears today evident that the present situation is quite different if we consider a *system* of reactions. There is no doubt that the molecules cannot migrate *against* a concentration gradient, but this becomes feasible if the process is coupled to chemical exergonic reaction. Alternatively, an isolated endergonic reaction cannot take place alone. Such reaction, however, can occur if coupled to the vectorial migration of molecules. In fact most of the endergonic reactions take place in the cell because they are coupled to other exergonic reactions, or to the vectorial transport of molecules. It is because the individual enzyme reactions are connected and form a network that individual disfavoured reactions take place. Put in other words, it is the network, the system, that allows the occurrence of disfavoured reactions. These simple considerations are consistent with the view that the living cell can be considered a complex system.

REFERENCES

[1]   Westerhoff, H. V. and Van Dam, K. (1987) Thermodynamics and Control of Biological Free Energy Transduction. Elsevier, Amsterdam.

[2]   Hill, T.L. (1968) Thermodynamics for Chemists and Biologists. Addison-Westley, Reading, MA.

[3]   Hill, T. L. (1977) Free Energy Transduction in Biology. Academic Press, New York.

[4]   Klipp, E., Libermeister, W., Wierling, C., Kowald, A., Lehrach, H. and Herwig, R. (2009) Systems Biology, Wiley-VCH Verlag, Weinheim.

[5]   Heinz, E. (1978) Mechanics and Energetics of Biological Transport. Springer Verlag, Berlin.

[6]   Heinz, E. (1981) Electrical Potentials in Biological Membrane Transport. Springer Verlag Berlin.

[7]   Cha, S. (1968) A simple method for derivation of rate equations under the rapid equilibrium assumption or combined assumption of equilibrium and steady state. J. Biol. Chem. 243, 820-825.

[8]   Martonosi, A.N. (1985) The Enzymes of Biological Membrane. Vol. 3. Membrane Transport, 2nd edition, Plenum Press, New York.

[9]   Stein, W. D. (1990) Carriers, Channels and Pumps: An Introduction of Membrane Transport. Academic Press, San Diego CA.

[10]   Ricard, J. (1999) Biological Complexity and the Dynamics of Life Processes. Elsevier, Amsterdam, Lausanne, New York.

[11]   Tanford, C. (1989) Mechanism of free energy coupling in active transport. Annu. Rev.Biochem. 52, 379-409.

[12]   Stein, W.D. (1986) Transport and Diffusion across Cell Membranes. Academic Press, Orlando. Fl.

[13]   Mercer, R. W. (1993) Structure of the Na-K ATPase. Int. Rev. Cytol. 137 C, 139-168.

[14]   Carafoli, E. (1991) Calcium pump of the plasma membrane. Physiol. Rev. 71, 129-153.

[15]   Hill, T. L. (1977) Free Energy Transduction in Biology. Academic Press, New York.

[16]   Boyer, P. D. (1977) The ATP synthase. A splendid molecular machine. Annu. Rev. Biochem. 66, 717-749.

[17]   Durand, R., Briand, Y., Touraille, S. and Alziari, S.(1981) Molecular approaches of phosphate transport in mitochondria. Trends Biochem. Sci. 6, 211-214.

[18]   Hinkle, P.C. and McCarty (1978) How cells make ATP. Sci. Amer. March, 104-123.

[19]   Mitchell. P. (1961) Coupling of phosphorylation to electron and hydrogen transfer by a chemi-osmotic type of mechanism. Nature 191, 144-148.

[20]   Hodgkin, A.L. and Keynes, R.D. (1955) Active transport of cations in giant axons from *Sepia* and *Loligo* J. Physiol. 164, 355-374.

[21]   Katz, B. (1966) Nerve, muscle and synapse. McGraw-Hill, New York.

[22]   Nicholls, D. G. (1982) Bioenergetics: An Introduction to the Chemiosmotic Theory Acad. Press, New York.

Send Orders for Reprints to reprints@benthamscience.net

CHAPTER 7

Driving Unfavorable Molecular Motions within the Cell

Abstract: Molecular motions take place within the cell. Considered in isolation these motions would have been impossible. It is only because they are coupled with chemical reactions within the cell that they can take place.

Keywords: Molecular motions within the cell, Tubulin, Microtubules, Actin, Dark bands, Light bands, Myosin, Treadmilling, Treadmill flow, Plus-end of a polymer, Minus-end of a polymer.

Organelles of the living cell are not inert, motionless, structures. They may appear, disappear, grow, shrink, bend, slide, and may therefore display different types of shape. All these changes appear spontaneous and one may raise the point of the origin(s) of these different types of morphological changes. We shall take as an example the case of microtubules and actin filaments which are the components of the cytoskeleton of eukaryotic cells.

1- TUBULIN AND MICROTUBULES

Tubulin is a heterodimer formed by two polypeptides called $\alpha-$ and $\beta-$tubulin. This protein dimer is present in virtually all eukaryotic cells. Tubulin can undergo spontaneous polymerization as to form microtubules. A microtubule is a hollow tube made up of 13 protofilaments each composed of alternating $\alpha-$ and $\beta-$ tubulin. Microtubules are not permanent entities of the cell. They are growing, or shrinking thanks to their polymerization, or depolymerization. One end of a microtubule, called "plus" grows thanks to tubulin polymerization. The other end, called "minus", shrinks owing to the dismantling of the supramolecular structure. The polymerization-depolymerization process is controlled by guanosine triphosphate (GTP) hydrolysis and exchange. This effect can be understood from the fact that GTP can bind to a site located on both $\alpha-$ and $\beta-$tubulins. It is interesting to note that whereas GTP bound to $\alpha-$tubulin cannot be hydrolysed or exchanged, GTP bound to $\alpha-$tubulin can be hydrolyzed to GDP (guanosine diphosphate). Moreover this GDP can be exchanged for another molecule of GTP present in the external milieu. *In vitro* tubulin can undergo polymerization in the presence of metal ions and GTP.

Jacques Ricard

Before polymerization starts, there is a nucleation phase corresponding to a lag of the polymerization process. The nucleation is followed by an elongation phase and finally by a steady state where the size of microtubules does not vary significantly. Under steady state conditions the polymerization process is such that the rate of tubulin binding at one end of the polymer (called plus) is equal to the rate of the tubulin release at the other end (called minus) of the microtubule (Fig. **1**). Such a situation means that the binding of a tubulin molecule to another one is a less probable event than the binding of a tubulin dimer to a microtubule. The elongation of microtubules takes place as long as GTP is bound to a specific site, called β, for tubulin units. If GTP is hydrolysed the microtubule disassembles.

In vivo, microtubules originate from a specialized cell structure called the centrosome. It appears then that the centrosome is the site of nucleation for microtubules that elongate in the cytoplasm. The mitotic spindle that appears during cell division consists of a set of microtubules [1-9].

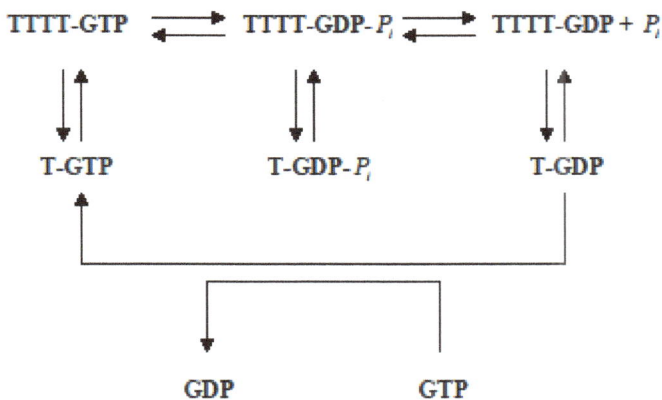

TTTT-GTP ⇌ TTTT-GDP-P_i ⇌ TTTT-GDP + P_i

T-GTP   T-GDP-P_i T-GDP

GDP   GTP

Figure 1: Nucleation phase of tubulin polymerization. See text.

2- Actin [9-20]

Actin (G-actin) is a protein present in all eukaryotic cells. It is made up of one polypeptide chain of 375 aminoacids. G-actin polymerizes as a left-handed helix called actin filament. Like microtubules, actin filaments are polymerized which means that the two ends are not equivalent. The plus end is a fast-growing structure whereas the minus end tends to disassemble. Moreover, the plus end has an arrowhead whereas the minus end has a barbed appearance. Actin filaments are often associated with myosin and form bundles that give the animal cells their shape. Actin polymerizes, following the same type of mechanism previously described for tubulin, with the exception that GTP is replaced by ATP. Any actin filament may exist in one of the following states: a F-ADP-state, a F-ADP-Pi-state, or a F-ATP-state (Fig. **2**). Even though ADP can be exchanged for ATP, this exchange is very slow. Exactly as for tubulin, actin polymerization requires mono- or divalent cations.

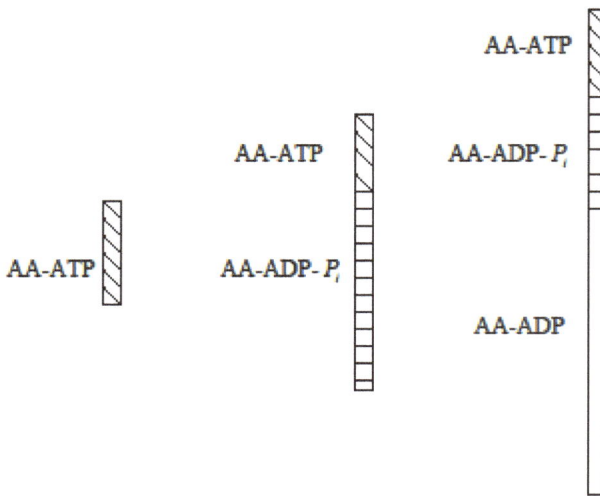

Figure 2: Different steps of actin elongation.

Large amounts of actin are present in muscle. Fibers of skeletal muscle are very large cells formed by the fusion of many different separate cells. The nuclei are located close to the plasma membrane and most of the cytoplasm is made up of myofibrils. Each myofibril consists of a sequence of elements called sarcomers that represent the machinery involved in muscle contraction. The myofibrils display dark and light bands. Located in the middle of each light band is a dense line called the Z line, or Z disk. The region limited by two Z disks is called a sarcomere. Actin filaments occupy the light band and part of the dark bands. Motor proteins, called myosins are also present in skeletal muscles. They possess two heavy chains and four light chains.The N-termini of the two heavy chains are called the globular heads that bear actin binding sites. The walk of myosin along an actin molecule requires ATP consumption. At the beginning of this process, a myosin head is tightly bound to an actin filament. Then an ATP molecule binds to the myosin head and this reduces the affinity of myosin for actin. ATP is then hydrolysed and its hydrolysis produces a conformational rearrangement of the myosin head that slides away from its initial position on actin. During the next step, the myosin head binds to a new region of actin and releases its ADP. This process generates a force exerted on actin and a sliding relative to myosin.

Let us consider the situation where a linear polymer of actin, an actin filament $A_D - A_D - A_D - ......$ for instance, has bound ADP on every actin unit. The polymer is assumed to be in equilibrium with monomers A_D that can be bound, or released, to and from the $\alpha(+)$ and $\beta(-)$ of the polymer. The flow, J_α, of A_D molecules that bind the $\alpha(+)$ end is

$$J_\alpha = \alpha_1 c_T - \alpha_{-1} \tag{7.1}$$

Similarly, the flow originating from the $\beta(+)$ end is

$$J_\beta = \beta_1 c_T - \beta_{-1} \tag{7.2}$$

In these expressions, c_T is the total concentration of monomers in solution. As the two ends, $\alpha(+)$ and $\beta(-)$, are different and one has $\alpha_1 \neq \beta_1$ and $\alpha_{-1} \neq \beta_{-1}$. Moreover the plus end is the site of polymerization. It follows that $\alpha_1 > \beta_1$ and $\alpha_{-1} < \beta_{-1}$. Hence, plotting J_α or J_β as a function of c_T should give straight plots (Fig. **3**).

Expressions (7.1) and (7.2) show that J_α and, or, J_β should be positive or negative depending on the concentration c_T. Moreover, there should exist a concentration, $\bar{c}_T$, where the forward and backward flows are equal. Let us consider an actin filament (A_D polymer in the Figure below). Let us assume that free actin molecules bind at both ends of the polymer. Let us consider that the free actin, A, and another actin molecule that has bound a colchicine molecule, AM, bind to the α – end and β – end of the polymer, respectively (Fig. **4**). In this Figure, c_A is the concentration of the free monomer and c_T its total concentration

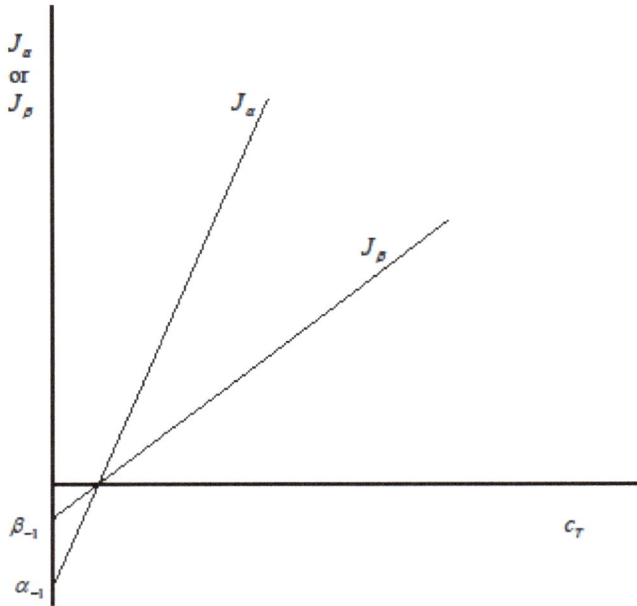

Figure 3: Polymerization flow of actin. Actin polymerization takes place at the two ends (α and β) of the polymer at different rates.

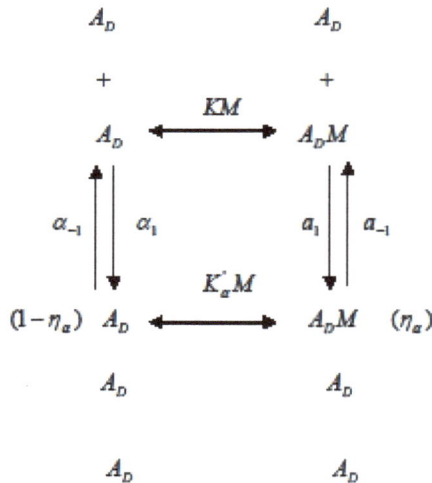

Figure 4: Drug binding on an equilibrium polymer. See text.

Under these conditions one has

$$c_A(1+Kc_M)=c_T.$$

(7.3)

which implies that

$$c_A = \frac{c_T}{1 + Kc_M} \tag{7.4}$$

Moreover the concentration of monomer, c_{AM}, that has bound colchicine is

$$c_{AM} = Kc_M c_A = \frac{Kc_M c_T}{1 + Kc_M} \tag{7.5}$$

The fraction of the liganded monomer at the $\alpha-$end of the polymer is

$$\eta_\alpha = \frac{K_\alpha' c_M}{1 + K_\alpha' c_M} \tag{7.6}$$

The fractions of $\alpha-$ and $\beta-$ends that have not bound M are thus $1 - \eta_\alpha$ and $1 - \eta_\beta$, respectively. It follows that the net flow of aggregation of the $\alpha-$end is then

$$J_\alpha = \alpha_1 c_A (1 - \eta_\alpha) + a_1 c_{AM} \eta_\alpha - \alpha_{-1}(1 - \eta_\alpha) - a_{-1}\eta_\alpha \tag{7.7}$$

It appears from this equation that under equilibrium conditions one has

$$\alpha_1 \overline{c}_A (1 - \eta_\alpha) - \alpha_{-1}(1 - \eta_\alpha) = 0 \tag{7.8}$$

and

$$a_1 \overline{c}_{AM} \eta_\alpha - a_{-1}\eta_\alpha = 0 \tag{7.9}$$

It follows from these equilibrium conditions that

$$\overline{c}_A = \frac{\alpha_{-1}}{\alpha_1} \text{ and } \overline{c}_{AM} = \frac{a_{-1}}{a_1} \tag{7.10}$$

These ratios are in fact identical to dissociation constants of unliganded and liganded subunits from the $\alpha-$end of the polymer.

There appears at this stage a contradiction between different results. This contradiction is only apparent. As the rate constant a_1 is very small one could conclude that the second term in equation (7.7) could be neglected. One could notice, however, that a_1 cannot be suppressed in equation (7.9). The explanation of this apparent contradiction is that the equilibrium of the polymer that has bound the colchicine at both ends should be shifted towards free subunits and hence should disassemble. Moreover one can notice that α_1 / α_{-1} and a_1 / a_{-1} cannot be independent for thermodynamics requires that (Fig. **4**).

$$K\left(\frac{\alpha_{-1}}{\alpha_1}\right) = K'_\alpha\left(\frac{a_{-1}}{a_1}\right) \tag{7.11}$$

This expression can be rewritten as

$$\frac{1}{K}\frac{\alpha_1}{\alpha_{-1}} = \frac{1}{K'_\alpha}\frac{a_1}{a_{-1}} \tag{7.12}$$

Experimental studies show that

$$\frac{\alpha_1}{\alpha_{-1}} > \frac{a_1}{a_{-1}} \tag{7.13}$$

which implies that

$$\frac{1}{K} < \frac{1}{K'_\alpha} \tag{7.14}$$

or

$$K > K'_\alpha \tag{7.15}$$

It then appears that the affinity of M for the free subunits is larger than for the $\alpha-$ end of the polymer. One can give a more conventional form to equation (7.7) if one takes account of the expressions of c_A, c_{AM} and η_α. One then obtains

$$J = \frac{c_T(\alpha_1 + a_1 KK'_\alpha c_M^2)}{(1 + Kc_M)(1 + K'_\alpha c_M)} - \frac{\alpha_{-1} + a_{-1}K'_\alpha c_M}{1 + K'_\alpha c_M} \tag{7.16}$$

As previously, the plots $J_\alpha = f(c_T)$ intersect the x axis and their slope decreases as c_M increases (Fig. **5**).

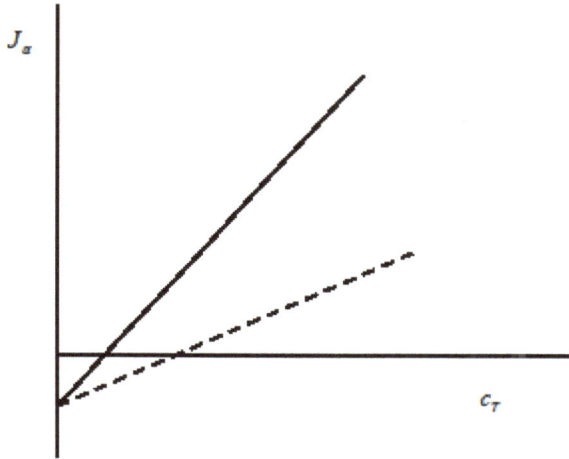

Figure 5: Inhibition of the polymerization flow of an equilibrium polymer. Colchicine has a stronger affinity for the free subunits than for the α -end of the polymer. This results in a decrease of the slope of the polymerization curve.

3- TREADMILLING

The growth of a polymer such as F-actin may take place under steady state. This implies that the monomers that have bound ATP (F-actin) or GTP (tubulin) bind to the $\alpha-$ or to the $\beta-$ end of the polymer. These monomers will be represented below by A_T . If the subunits of the polymer are bearing ADP or GDP they will be termed. A_D . If an A_T monomer binds to the $\alpha-$, or to the $\beta-$ end of a polymer it undergoes the hydrolysis of its ATP. As we have already pointed out, the polymer is polarized, which means that the properties of the two ends are different. The different properties of the two ends of the polymer explain that any G-actin unit that binds to the polymer moves along the macromolecule (Fig. **6**). This phenomenon is called treadmilling.

In this process one end called "plus" is polymerized and grows whereas the other end, called "minus", depolymerises. The kinetics and thermodynamics of this process can be treated in the following way. At the two ends of actin filaments two events take place. The first one is the association of an actin monomer bearing ATP, and called below A_T, with the $\alpha-$end of the polymer. During the polymerization process, ATP is being converted into ADP (D) and phosphate (P). One has then

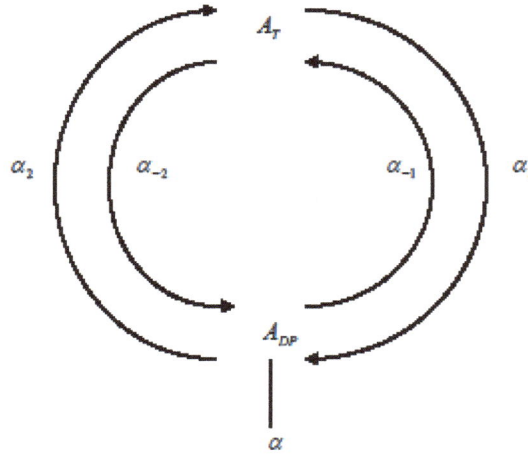

Figure 6: Kinetic scheme of the polymerization process at the α -end of a steady state polymer. See text.

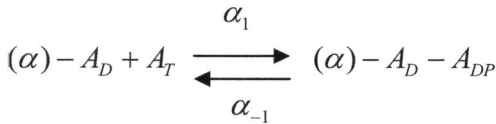

$$(\alpha)-A_D+A_T \xrightleftharpoons[\alpha_{-1}]{\alpha_1} (\alpha)-A_D-A_{DP}$$

The same process is taking place, but at different rate, at the $\beta-$end of the actin filament.

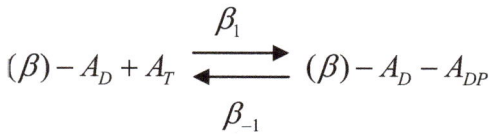

$$(\beta)-A_D+A_T \xrightleftharpoons[\beta_{-1}]{\beta_1} (\beta)-A_D-A_{DP}$$

The rate constants α_1 and α_{-1}, β_1 and β_{-1} are indeed different but if the actin filament is in steady state, which implies that its rate of elongation is constant, then one has

$$\frac{\alpha_1}{\alpha_{-1}}=\frac{\beta_1}{\beta_{-1}} \tag{7.17}$$

The second type of process that takes place at the two ends of the polymer is the release of actin unit from the polymer together with the release of ADP (D) and phosphate (P) and their replacement by an ATP (T). One has

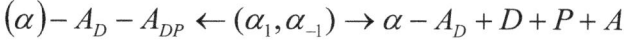

$$(\alpha) - A_D - A_{DP} \leftarrow (\alpha_1, \alpha_{-1}) \rightarrow \alpha - A_D + D + P + A$$

$$A + T \leftarrow\rightarrow AT$$

and at the $\beta-$ end

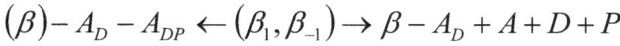

$$(\beta) - A_D - A_{DP} \leftarrow (\beta_1, \beta_{-1}) \rightarrow \beta - A_D + A + D + P$$

It is evident that the rate constants α_1 and α_{-1} should be different from β_1 and β_{-1} but the two equilibrium constants should be the same. Hence one has

$$\frac{\alpha_1}{\alpha_{-1}} = \frac{\beta_1}{\beta_{-1}} \tag{7.18}$$

Other events are also taking place at the two ends of the polymer. As a matter of fact, ATP (T) is synthesized from ADP (D) and phosphate (P). Hence one has

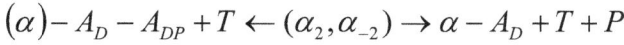

$$(\alpha) - A_D - A_{DP} + T \leftarrow (\alpha_2, \alpha_{-2}) \rightarrow \alpha - A_D + T + P$$

and

$$(\beta) - A_D - A_{DP} + T \leftarrow (\beta_2, \beta_{-2}) \rightarrow \beta - A_D + T + P$$

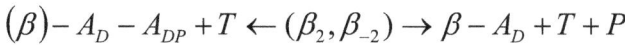

Here again one should expect that

$$\frac{\alpha_2}{\alpha_{-2}} = \frac{\beta_2}{\beta_{-2}} \tag{7.19}$$

Let us call c_{AT} the concentration of the A_T monomer, it follows that the rate of addition of monomers at the two ends, α and β, of the same polymer are

$$\frac{dn_\alpha}{dt} = (\alpha_1 + \alpha_{-2}) c_{AT} - (\alpha_{-1} + \alpha_2) \tag{7.20}$$

$$\frac{dn_\beta}{dt} = (\beta_1 + \beta_{-2})c_{AT} - (\beta_{-1} + \beta_2) \tag{7.21}$$

The rate of motion of a monomer along the polymer, *i.e.* the treadmill flow $\tilde{J}$ is then

$$\tilde{J} = \frac{dn_\alpha}{dt} = -\frac{dn_\beta}{dt} \tag{7.22}$$

Under the conditions of treadmilling the steady concentration of the monomer, c_{AT}, is such that

$$(\alpha_1 + \alpha_{-2})\overline{c}_{AT} - (\alpha_{-1} + \alpha_2) = -(\beta_1 + \beta_{-2})\overline{c}_{AT} + (\beta_{-1} + \beta_2) \tag{7.23}$$

or

$$\overline{c}_{AT} = \frac{\alpha_2 + \beta_2 + \alpha_{-1} + \beta_{-1}}{\alpha_1 + \beta_1 + \alpha_{-2} + \beta_{-2}} \tag{7.24}$$

The expression of the treadmill flow can be expressed as

$$\tilde{J} = \frac{\alpha_1\beta_{-1} - \alpha_{-1}\beta_1 + \alpha_{-2}\beta_2 - \alpha_2\beta_{-2} + \alpha_1\beta_2 - \alpha_2\beta_1 + \alpha_{-2}\beta_{-1} - \alpha_{-1}\beta_{-2}}{\alpha_1 + \beta_1 + \alpha_{-2} + \beta_{-2}} \tag{7.25}$$

Moreover one can write from expressions (7.19) and (7.20)

$$\alpha_1\beta_{-1} = \alpha_{-1}\beta_1$$

$$\alpha_2\beta_{-2} = \alpha_{-2}\beta_2 \tag{7.26}$$

$$\frac{\alpha_1\alpha_2}{\alpha_{-1}\alpha_{-2}} = \frac{\beta_1\beta_2}{\beta_{-1}\beta_{-2}}$$

and expression (7.25) can then be rewritten as

$$\tilde{J} = \frac{(\alpha_1\beta_2 - \alpha_2\beta_1)\left(1 - \dfrac{\alpha_{-1}\alpha_{-2}}{\alpha_1\alpha_2}\right)}{\alpha_1 + \beta_1 + \alpha_{-2} + \beta_{-2}} \tag{7.27}$$

One can notice that

$$\frac{\alpha_1\alpha_2}{\alpha_{-1}\alpha_{-2}} = \frac{\beta_1\beta_2}{\beta_{-1}\beta_{-2}} = \exp\left(\frac{\mu_T - \mu_D - \mu_P}{RT}\right) \tag{7.28}$$

where μ_T, μ_D and μ_P are the chemical potentials of ATP, ADP and phosphate, respectively. One can also notice that

$$X = \mu_T - \mu_D - \mu_P \tag{7.29}$$

is the driving force of the treadmill flow. It then appears that the treadmill flow can be expressed in thermodynamic terms, namely

$$\tilde{J} = \frac{(\alpha_1\beta_2 - \alpha_2\beta_1)(1 - e^{-X/RT})}{\alpha_1 + \beta_1 + \alpha_{-2} + \beta_{-2}} \tag{7.30}$$

This equation has two important implications. It shows that the existence of treadmilling relies upon the force X. If this force is nil no treadmilling is expected to occur. Moreover if $\alpha_1 = \beta_1$ and $\alpha_2 = \beta_2$ the polymer is not polarized and again no treadmilling can take place. It is noteworthy that it is nonequilibrium thermodynamics that offers the proof that treadmilling relies upon the polarization of actin filaments as well as upon the consumption of energy.

One can notice that the conversion of A_T to $(\alpha) - A_D - A_{DP}$ and back constitutes a cycle shown in Fig. **7**. Hence it is of interest to wonder how much energy is required to keep this cycle running. One has for the first part of the cycle

$$\mu_{AT} = \mu_{AT}^\circ + RT \ln \overline{c}_{AT} = \mu_{ADP} \tag{7.31}$$

In this expression $\overline{c}_{AT}$ is the equilibrium concentration of A_T which is itself equal to α_{-1}/α_1. Hence one has

$$RT \ln \frac{\alpha_1}{\alpha_{-1}} = \mu_{AT}^\circ - \mu_{ADP} \tag{7.32}$$

Moreover one can also write

$$\mu_{ADP} + \mu_T = \mu_{AT} + \mu_D + \mu_P \tag{7.33}$$

and for global equilibrium conditions

$$\mu_{AT} = \mu_{AT}^\circ + RT \ln \overline{c}_T = \mu_{AT}^\circ + RT \ln \frac{\alpha_2}{\alpha_{-2}} \tag{7.34}$$

Hence equation (7.33) can be rewritten as

$$\mu_{ADP} + \mu_T = \mu_A^\circ + RT \ln \overline{c}_T + \mu_D + \mu_P \tag{7.35}$$

and this expression is equivalent to

$$\mu_{ADP} + \mu_T = \mu_{AT}^\circ + RT \ln \frac{\alpha_2}{\alpha_{-2}} + \mu_D + \mu_P \tag{7.36}$$

or to

$$RT \ln \frac{\alpha_2}{\alpha_{-2}} = \mu_{ADP} + \mu_T - \mu_D - \mu_P - \mu_{AT}^\circ \tag{7.37}$$

and

$$RT \ln \frac{\alpha_2}{\alpha_{-2}} = \mu_{ADP} + X - \mu_{AT}^\circ \tag{7.38}$$

where X is still the force that drives the treadmill flow (Fig. **7**). One can notice that the energy required to keep the cycle running is

$$RT \ln \frac{\alpha_1}{\alpha_{-1}} + RT \ln \frac{\alpha_2}{\alpha_{-2}} = RT \ln \frac{\alpha_1 \alpha_2}{\alpha_{-1} \alpha_{-2}} = X \tag{7.39}$$

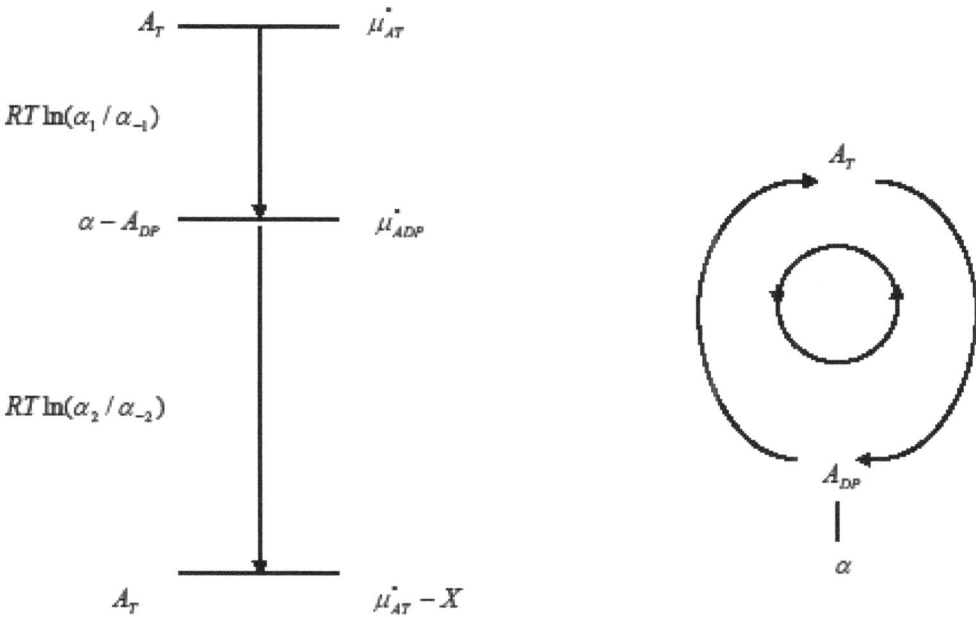

Figure 7: Energetics of the A_T, ADP cycle. Right: The $A_T, \alpha - ADP$ cycle. Left: Energetics of the cycle.

4- GENERAL CONCLUSIONS

It may appear surprising that several supramolecular edifices present in the living cell are continuously built up and dismantled. This is the case for instance for the mitotic spindle that appears during cell division and disappears afterwards. A similar situation occurs for actin filaments that grow and shrink depending on the experimental conditions. The physical reasons for this situation are simple. On the one hand, these edifices are polarized, *viz.* they possess a plus and a minus end. Moreover, these polarized structures do not occur under equilibrium conditions. This means that some energy is being continuously spent in order to avoid shrinkage and dismantlement. The energy originates from GTP or ATP hydrolysis depending on the filament. The process of treadmilling, *i.e.* the motion of actin molecules along an actin filament can be explained in physical terms only through a coupling of the supramolecular edifice with ATP consumption. The basic idea that can be drawn from these studies is that the dynamics of the intracellular structure is linked to the hydrolysis of substances such as ATP or GTP.

REFERENCES

[1]   Dustin, P. (1984) Microtubules, 2nd Edition, Springer-Verlag, New York.

[2]   Amos, L. A. and Baker, T. S. (1979) The three-dimensional structure of tubulin protofilaments. Nature 279, 607-612.

[3]   Sullivan, K. F. (1988) Structure and utilization of tubulin isotypes Annu. Rev. Cell Biol. 4, 687-716.

[4]   De Brabander, M. (1986) Microtubule dynamics during the cell cycle: the effect of taxol and nocodazole on the microtubular system of $[C_1]/[C_0],[C_2]/[C_0]$ cells at different stages of the mitotic cycle. Int. Rev. Cytol. 101, 215-274.

[5]   Bergen, L. G. and Borisy, G.G. (1980) Head-to-tail polymerization of microtubules *in vitro*. Electron microscope analysis of seeded assembly. J. Cell Biol. 84, 141-150.

[6]   McIntosh, J. R. and Euteneur, U. (1984) Tubulin hooks as probes for microtubule polarity: an analysis of the method of evaluation of the data on microtubule polarity in the mitotic spindle. J. Cell Biol.88, 525-533.

[7]   Mitchinson, T. J. and Kirschner, M.W. (1984) Dynamic instability of microtubule growth. Nature 312, 237-242.

[8]   Olmsted, J. B. (1986) Microtubule-associated proteins. Annu. Rev. Cell Biol.2, 421-457.

[9]   Frieden, C. (1985) Actin and tubulin polymerization. The use of kinetic methods to determine mechanism. Annu. Rev. Biophys. Biophys. Chem. 14, 189-210.

[10]   Pollard, T. D. and Cooper, J. A. (1986) Actin and actin-binding proteins. Acritical evaluation of mechanisms and functions. Annu. Rev. Biochem. 55, 987-1035.

[11]   Carlier, M. F., Pantaloni, D. and Korn, E. D. (1986) Fluorescence measurements of the binding of cations to high-affinity and low-affinity sites of G-actin. J. Biol. Chem. 261, 10778-10784.

[12]   Estes, J. E., Selden, L. A. and Gershman, L. C. (1987) Tight binding of divalent cations to monomeric actin. Binding kinetics supports a simplified model. J. Biol. Chem. 262, 4952-4957.

[13]   Carlier, M. F. Pantaloni, D. and Korn, E. D. (1986) The effects of Mg at the high affinity and low affinity sites on the polymerization of actin and associated ATP hydrolysis. J. Biol. Chem. 261, 10785-10792.

[14]   Hill, T. L. and Kirschner, M. (1983) Regulation of microtubule and actin filament assembly-disassembly by associated small and large molecules. Int. Rev. Cytol. 84, 185-234.

[15]   Hill, T. L. (1981) Steady state head-to-tail polymerization of actin or microtubules. II. Two-state and three states kinetic cycles. Biophys. J. 33, 353-371.

[16]   Hill, T. L. (1980) Bioenergetic aspects and polymer distribution in steady state and head-to-tail polymerization of actin or microtubules. Proc. Natl. Acad. Sci. USA 77, 4803-4807.

[17]   Hill, T. L. (1977) Free Energy Transduction in Biology. Academic Press, New York.

[18]   Julicher, F. and Prost, J. (1995) Cooperative molecular motors. Phys. Rev. Letters 75, 421-426.

[19]   Julicher, F. and Prost, J. (1997) Spontaneous oscillations of collective molecular motors. Phys. Rev. Lett., 78, 4510-4513.

[20]   Ricard, J. (1999) Biological Complexity and the Dynamics of Life Processes. Elsevier, Amsterdam, Lausanne, New York.

Send Orders for Reprints to reprints@benthamscience.net

CHAPTER 8

Physics of Metabolic Oscillations

Abstract: Individual enzyme reactions in isolation cannot display any periodic behavior, however they display this behavior when they are associated with other reactions. It is the system thus formed that displays such periodic behavior.

Keywords: Oscillations of a metabolic cycle, Dimensionless variables, Dynamics and stability of a model cycle, Linear variational system, Trace and determinant of a jacobian matrix, Temporal organization at the surface of a charged membrane, Partition coefficient, Oscillations of a partition coefficients, Jacobian matrix, Trace and determinant of the jacobian matrix, Multistability of a chemical system.

Individual enzyme reactions considered in isolation do not display periodic behaviour. However, such periodic behavior exists at a more global level. This is the case for glycolysis that displays such oscillations *in vivo*. Morever, surprising effects have been observed in yeast cells and in cell free extracts [1-8]. These effects can be either relatively simple, or more complex as they may display, depending on the experimental conditions, bi-periodicity or chaos. Another phenomenon displaying periodic behavior is known under the name of calcium spiking. It can be observed in some animal cells such as hepatocytes and endothelial cells. We intend, in this Chapter, to describe in mathematical terms the rationale for this periodic, or aperiodic behavior [9-16].

1- CONDITIONS OF EMERGENCE OF OSCILLATIONS IN A MODEL METABOLIC CYCLE

Let us consider the simplest open metabolic cycle (Fig. **1**). It is made up of two antagonistic enzyme reactions, an input and an output of metabolites. The two reagents are S_1 and S_2. v_1 and v_2 are the two reaction rates of enzyme reactions. v_i and v_0 are the rates of input and output of matter.

The corresponding variational system can be written as

$$\frac{dS_1}{dt} = v_i - v_1 + v_2 \qquad (8.1)$$

$$\frac{dS_2}{dt} = v_1 - v_2 - v_0$$

where v_1 and v_2 are

$$v_1 = \frac{V_1 S_1}{K_1 + S_1} \qquad\qquad (8.2)$$

$$v_2 = \frac{V_2 S_2}{K_2 + S_2}$$

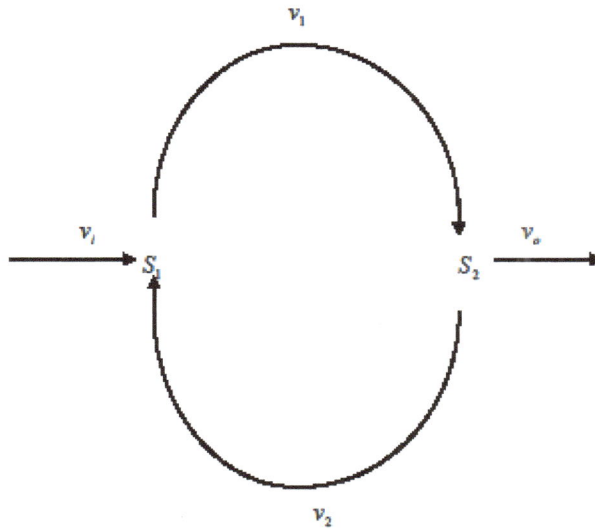

Figure 1: An open metabolic cycle. See text.

where V_1 and V_2 are the maximum reaction rates, K_1 and K_2 the corresponding Michaelis constants. Moreover v_i is the input rate of S_2 in the system and v_0 the output rate of S_2 from this system.

The system (8.1) above can be written in dimensionless form after having defined the dimensionless variables

$$\alpha = \frac{S_1}{K_1} \quad \beta = \frac{S_2}{K_2}$$

$$v_1 = \frac{v_1}{V_1} \quad v_2 = \frac{v_2}{V_1} \quad v_i = \frac{v_i}{V_1} \quad v_0 = \frac{v_0}{V_1} \tag{8.3}$$

$$\varepsilon = \frac{K_1}{K_2} \quad \theta = \frac{V_1}{K_1} t \quad \lambda = \frac{V_2}{V_1}$$

Moreover in equations (8.1) v_i is the input rate of S_1 in the system and v_0 the output rate of S_2 from this system. The system (8.1) above can be written in dimensionless form after having defined dimensionless variables (expressions 8.3 above). The dynamic system (8.1) can then be expressed in dimensionless form as

$$\frac{d}{d\theta}\begin{bmatrix} \alpha \\ \beta \end{bmatrix} = \begin{bmatrix} 1 & 0 \\ 0 & \varepsilon \end{bmatrix}\begin{bmatrix} v_i - v_1 + v_2 \\ v_1 - v_2 - v_0 \end{bmatrix} \tag{8.4}$$

whereas the original dynamic system (equations 8.1 and 8.2) contains nine parameters the same normalized system possesses only seven independent parameters.

Now let us consider the two functions

$$v_i - v_1 + v_2 = v_i - v_1(\alpha,\beta) + v_2(\alpha,\beta) = u_1(\alpha,\beta) \tag{8.5}$$

$$v_1 - v_2 - v_0 = v_1(\alpha,\beta) - v_2(\alpha,\beta) - v_0 = u_2(\alpha,\beta)$$

In the second equation (8.5) v_0 is first order in S_2 (or β) and one has

$$v_0 = \frac{kK_2}{V_1}\beta = \mu\beta \tag{8.6}$$

If the system is in steady state equations (8.5) and (8.6) above become

$$v_i - v_1(\alpha,\beta) + v_2(\alpha,\beta) = 0$$
$$v_1(\alpha,\beta) - v_2(\alpha,\beta) - \mu\beta = 0 \tag{8.7}$$

The first of these equations can be rewritten as

$$v_i = v_1(\alpha, \beta) - v_2(\alpha, \beta) \tag{8.8}$$

and from the second equation (8.7)

$$\mu\beta = v_1(\alpha, \beta) - v_2(\alpha, \beta) \tag{8.9}$$

comparing equations (8.8) and (8.9) yields

$$\beta = \frac{v_i}{\mu} \tag{8.10}$$

The steady state concentration α^* is then a solution of

$$v_1(\alpha, \beta^*) - v_2(\alpha, \beta^*) - \mu\beta^* = 0 \tag{8.11}$$

Moreover equation (8.10) shows there is one steady state value β, but equation (8.11) leaves it open the possibility of several steady states for α. In fact, multiple steady states rely upon the individual reaction rates. If, for instance, the two enzymes follow Michaelis-Menten kinetics one has, in dimensionless form

$$v_1 = \frac{\alpha}{1+\alpha} \tag{8.12}$$

$$v_2 = \frac{\lambda\beta}{1+\beta}$$

and equation (8.11) above becomes

$$\frac{\alpha}{1+\alpha} - \frac{\lambda\beta^*}{1+\beta^*} - \mu\beta^* = 0 \tag{8.13}$$

The steady state expression of α, α^*, can be derived from this equation and one finds

$$\alpha^* = \frac{\lambda\beta^*/(1+\beta^*) + \mu\beta^*}{1 - \lambda\beta^*/(1+\beta^*) - \mu\beta^*} \tag{8.14}$$

It appears from this equation that a steady state for α exists only if

$$1 > \mu\beta^* + \frac{\lambda\beta^*}{1+\beta^*} \tag{8.15}$$

Moreover one can observe from equation (8.13) that a steady state of the system relative to β^* is obtained if

$$\frac{\alpha}{1+\alpha} = \mu\beta^* + \frac{\lambda\beta^*}{1+\beta^*} \tag{8.16}$$

Setting

$$Y_1(\alpha) = \frac{\alpha}{1+\alpha} \tag{8.17}$$

and

$$Y_2(\beta^*) = \mu\beta^* + \frac{\lambda\beta^*}{1+\beta^*} \tag{8.18}$$

It appears that Y_1 is the rate of production of β and Y_2 the rate of consumption of α. Plotting Y_1 and Y_2 as a function of α yields a rectangular hyperbola and a straight line, respectively (Fig. **2**). Fig. **2** shows that there may exist one point of intersection between Y_1 and Y_2. This imply that the system should be stable, *i.e.* slightly perturbed from its steady state it comes back to the same steady state.

If, however, one of the two functions, for instance Y_2, is a function of both α and β it may occur that $Y_2(\alpha)$ intersects $Y_1(\alpha)$ at two points. One of these intersections would then be stable and the other unstable.

2- DYNAMICS AND STABILITY ANALYSIS OF A MODEL CYCLE

Let us consider the simple model cycle of Fig. **1**. The time evolution of normalized concentrations α and β are expressed as

$$\frac{d\alpha}{dt} = \frac{d\alpha^*}{dt} + \frac{dx_\alpha}{dt} \tag{8.19}$$

$$\frac{d\beta}{dt} = \frac{d\beta^*}{dt} + \frac{dx_\beta}{dt}$$

where x_α and x_β are small deviations relative to steady state concentrations α^* and β^*. It follows that

$$\frac{d\alpha}{dt} = \frac{dx_\alpha}{dt} \tag{8.20}$$

$$\frac{d\beta}{dt} = \frac{dx_\beta}{dt}$$

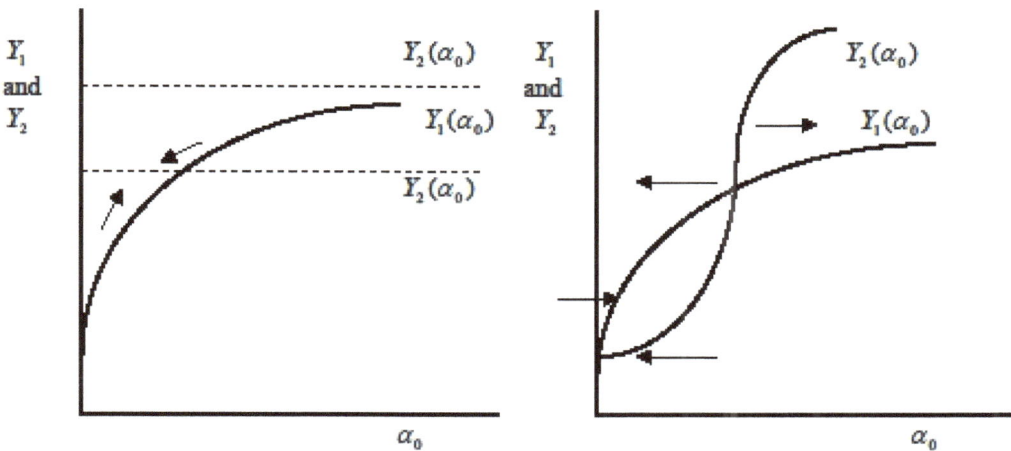

Figure 2: Stable and unstable steady states of the open cycle. Left: $Y_1(\alpha_0)$ is a hyperbola and $Y_2(\alpha_0)$ a straight line. The two functions can have one, or zero point of intersection. Right: The first function $Y_1(\alpha_0)$ is a function of α_0 only. The second function, $Y_2(\alpha_0, \beta_0)$ is a function of both α_0 and β_0. If α_0 is varied, Y_1 and Y_2 have two intersection points, one is stable, the other unstable.

As far as x_α and x_β are small relative to α and β, the system (8.19) can be studied trough the so-called linear variational system (system 8.21). One has

$$\frac{d}{dt}\begin{bmatrix} x_\alpha \\ x_\beta \end{bmatrix} = \begin{bmatrix} \partial u_1^*/\partial\alpha & \partial u_1^*/\partial\beta \\ \partial u_2^*/\partial\alpha & \partial u_2^*/\partial\beta \end{bmatrix}\begin{bmatrix} x_\alpha \\ x_\beta \end{bmatrix} \tag{8.21}$$

Here the u^*'s are defined by equations (8.5) and (8.6). The characteristic equation of this variational system is then

$$D^2 - T_j D + \Delta_j = 0 \tag{8.22}$$

where D is the differential operator d/dt, T_j and Δ_j the trace and the determinant of the jacobian matrix

$$J = \begin{bmatrix} \partial u_1^* / \partial \alpha & \partial u_1^* / \partial \beta \\ \partial u_2^* / \partial \alpha & \partial u_2^* / \partial \beta \end{bmatrix} \tag{8.23}$$

The trace and the determinant of this jacobian matrix are then

$$T_j = \frac{\partial u_1^*}{\partial \alpha} + \frac{\partial u_2^*}{\partial \beta} \tag{8.24}$$

$$\Delta_j = \frac{\partial u_1^*}{\partial \alpha} \frac{\partial u_2^*}{\partial \beta} - \frac{\partial u_2^*}{\partial \alpha} \frac{\partial u_1^*}{\partial \beta}$$

and the general solution of this system is

$$\begin{aligned} x_1 &= c_{11} \exp(\lambda_1 t) + c_{12} \exp(\lambda_2 t) \\ x_2 &= c_{21} \exp(\lambda_1 t) + c_{22} \exp(\lambda_2 t) \end{aligned} \tag{8.25}$$

The trace T_j and the determinant Δ_j can also be expressed from the two rate constants λ_1 and λ_2. One has

$$T_j = \lambda_1 + \lambda_2 \tag{8.26}$$

$$\Delta_j = \lambda_1 \lambda_2$$

It then follows that the dynamic behavior of the system is defined by the respective values of T_j, Δ_j and $T_j^2 - 4\Delta_j$.

The parabola $T_j^2 - 4\Delta_j = 0$ defines six regions in the (T_j, Δ_j) plane (Fig. 3). In region I one has

$$T_j < 0$$

$$\Delta_j > 0 \tag{8.27}$$

$$T_j^2 > 4\Delta_J$$

It follows that the two roots λ_1 and λ_2 are real and negative. This conclusion implies that the system is stable. If perturbed from its steady state it comes back to the same steady state. In region IV one has

$$T_J > 0$$

$$\Delta_J > 0 \tag{8.28}$$

$$T_J^2 > 4\Delta_J$$

The system is unstable because the two roots are real and positive. If perturbed from its steady state it monotonically drifts to a new steady state.

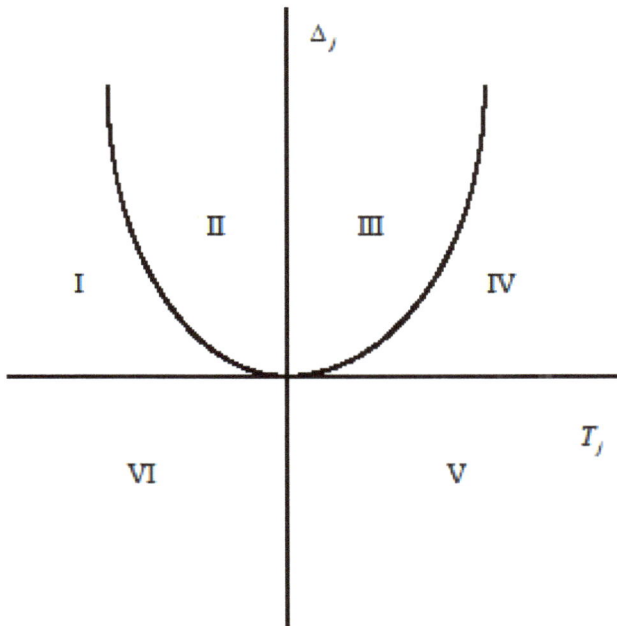

Figure 3: The $T_j - \Delta_j$ plane. See text.

In regions II and III the situation is totally different. For region II one has

$$T_J < 0$$

$$\Delta_J > 0 \tag{8.29}$$

$$4\Delta_J > T_J^2$$

These relationships imply that the two roots are complex with negative real parts. One has

$$\lambda_{1,2} = \frac{T_J}{2} \pm i\omega \tag{8.30}$$

$$\omega = \frac{1}{2}\sqrt{4\Delta_J - T_J^2}$$

The general solution of the system is then

$$x_1 = \exp(T_J t / 2)\{C_{11}\exp(i\omega t) + C_{12}\exp(-i\omega t)\} \tag{8.31}$$

$$x_2 = \exp(T_J t / 2)\{C_{21}\exp(i\omega t) + C_{22}\exp(-i\omega t)\}$$

As T_J is negative, the system is stable. Perturbed from its steady state, it returns back to the same steady state through damped oscillations. The system displays a stable focus.

For region III,

$$T_J > 0$$

$$\Delta_J > 0 \tag{8.32}$$

$$4\Delta_J > T_J^2$$

and the two roots are complex with positive real parts. Perturbed from its steady state, the system displays amplified oscillations and drifts towards a new steady state. The system has then an unstable focus.

In regions V and VI one has

$T_J > 0$ (region V), $T_J < 0$ (region VI)

$$\Delta_J < 0 \tag{8.33}$$

$$T_J^2 > 4\Delta_J$$

The two roots are real and opposite in sign. The system is unstable and displays a saddle point.

A particularly interesting situation is obtained if

$$T_J = 0$$

$$\Delta_J > 0 \tag{8.34}$$

$$4\Delta_J > T_J^2$$

Then the system displays sustained oscillations and a limit cycle (Fig. **4**). This is the situation which is occurring if the enzyme E_2 is inhibited by an excess substrate and follows an equation of the type

$$v_2 = \frac{\lambda\beta}{1 + \beta + \xi\beta^2} \tag{8.35}$$

An important conclusion that can be derived from the above theoretical results is that coupled "simple" enzymes that follow Michaelis-Menten kinetics cannot display sustained oscillations. The minimum required to obtain such a situation is that one of the reaction rates be inhibited by an excess substrate.

3- EMERGENCE OF A TEMPORAL ORGANIZATION AT THE SURFACE OF A CHARGED MEMBRANE

The above reasoning can be extended to a metabolic cycle located at the surface of a charged membrane. Let us consider the situation shown in Fig. **5**. Both an influx and an outflux of ions take place at the surface of a charged matrix [11-16]. Two enzymes convert S_1 into S_2, and back. One reaction takes place at the external surface of the matrix whereas the reverse process occurs inside the matrix (Fig. **5**). For simplicity, it is assumed that the two reactions are nearly irreversible. This may be achieved if the forward reaction requires the participation of a reagent X_1, and the backward process a different substance X_2, in such a way that $S_1 + X_1 \rightarrow S_2 + X_1'$ and $S_2 + X_2 \rightarrow S_1 + X_2'$. If X_1 and X_2 are saturating these two reactions are irreversible and the concentrations of X_1, X_1', X_2, X_2' can be cancelled out from the mathematical treatment of the dynamic process.

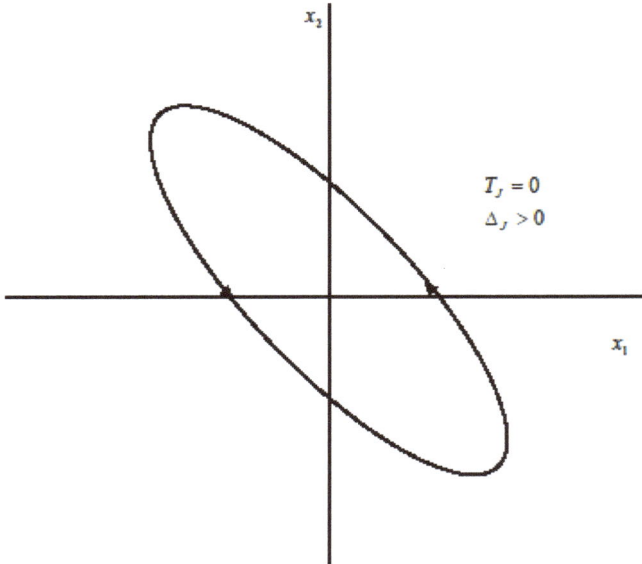

Figure 4: The limit cycle. See text.

One can define a partition coefficient, Π, from the ratios of the concentrations of anions, or cations, in the bulk and matrix phases. If Δ is the charge of the matrix, the partition coefficient of ions can be written as

$$\Pi = \frac{\sqrt{\Delta^2 + 4(S_{10} + S_{20})^2} + \Delta}{2(S_{10} + S_{20})} = \frac{2(S_{10} + S_{20})}{\sqrt{\Delta^2 + 4(S_{10} + S_{20})^2} - \Delta} \qquad (8.36)$$

In dilute solutions, the partition coefficient can be approximated to the ratio of the internal and external concentrations. In the case of the present model, one has

$$\Pi = \frac{H_i}{H_0} = \frac{S_{10}}{S_{1i}} = \frac{S_{20}}{S_{2i}} = \exp\left(\frac{F\Delta\Psi}{RT}\right) \tag{8.37}$$

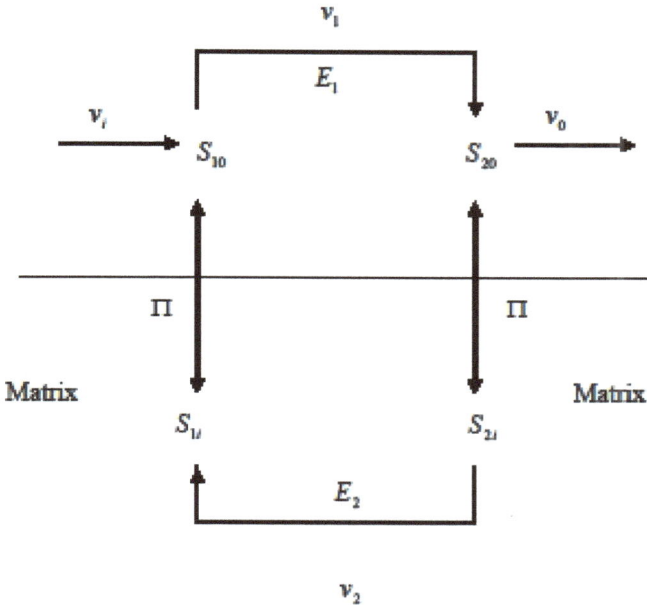

Figure 5: A simple open system at the surface of a charged matrix. The enzyme E_1 is responsible for the conversion $S_1 \to S_2$ outside the charged matrix. The enzyme E_2 is responsible for the conversion $S_2 \to S_1$ inside the matrix.

In equations (8.36) and (8.37) the subscripts i and o refer to the inside and the outside of the charged matrix. It is then easy to express the local concentration (concentration in the matrix) of ions as a function of the corresponding external concentration and of the charge Δ. One finds

$$S_{1i} = \frac{S_{10}}{2(S_{10}+S_{20})}\left\{\sqrt{\Delta^2+4(S_{10}+S_{20})^2} - \Delta\right\}$$

$$\tag{8.38}$$

$$S_{2i} = \frac{S_{20}}{2(S_{10}+S_{20})}\left\{\sqrt{\Delta^2+4(S_{10}+S_{20})^2} - \Delta\right\}$$

If one assumes that the equilibration of S_1 and S_2 in the two phases is fast relative to the other processes that may take place, one has

$$\frac{d(S_{10} + S_{1i})}{dt} = \frac{dS_{10}}{dt} + \frac{dS_{1i}}{dt} = v_i - v_1 + v_2 \tag{8.39}$$

$$\frac{d(S_{20} + S_{2i})}{dt} = \frac{dS_{20}}{dt} + \frac{dS_{2i}}{dt} = v_1 - v_2 - v_0$$

where v_i and v_0 are the rates of input and output in and from the system, respectively. Moreover one can write

$$\frac{dS_{1i}}{dt} = \left(\frac{\partial S_{1i}}{\partial S_{10}}\right)\frac{dS_{10}}{dt} + \left(\frac{\partial S_{1i}}{\partial S_{20}}\right)\frac{dS_{20}}{dt} \tag{8.40}$$

$$\frac{dS_{2i}}{dt} = \left(\frac{\partial S_{2i}}{\partial S_{10}}\right)\frac{dS_{10}}{dt} + \left(\frac{\partial S_{2i}}{\partial S_{20}}\right)\frac{dS_{20}}{dt}$$

and equations (8.39) can be rewritten as

$$\left\{1 + \left(\frac{\partial S_{1i}}{\partial S_{10}}\right)\right\}\frac{dS_{10}}{dt} + \left(\frac{\partial S_{1i}}{\partial S_{20}}\right)\frac{dS_{20}}{dt} = v_i - v_1 + v_2 \tag{8.41}$$

$$\left(\frac{\partial S_{2i}}{\partial S_{10}}\right)\frac{dS_{10}}{dt} + \left\{1 + \left(\frac{\partial S_{2i}}{\partial S_{20}}\right)\right\}\frac{dS_{20}}{dt} = v_1 - v_2 - v_0$$

These equations have to be compatible with the mass balance equation

$$\frac{dn_{10}}{dt} + \frac{dn_{20}}{dt} = 0 \tag{8.42}$$

where n_{10} and n_{20} are the mole numbers of reagents S_1 and S_2 outside the matrix. Expression (8.41) requires that $v_i = v_0$. From equations (8.41) one has

$$\left\{1 + \left(\frac{\partial S_{1i}}{\partial S_{10}}\right) + \left(\frac{\partial S_{2i}}{\partial S_{10}}\right)\right\}\frac{dS_{10}}{dt} = \left\{1 + \left(\frac{\partial S_{1i}}{\partial S_{20}}\right) + \left(\frac{\partial S_{2i}}{\partial S_{20}}\right)\right\}\frac{dS_{20}}{dt} \tag{8.43}$$

This equation requires that

$$\left(\frac{\partial S_{1i}}{\partial S_{10}}\right) + \left(\frac{\partial S_{2i}}{\partial S_{10}}\right) = \left(\frac{\partial S_{1i}}{\partial S_{20}}\right) + \left(\frac{\partial S_{2i}}{\partial S_{20}}\right) \tag{8.44}$$

One may notice that

$$\frac{\Delta}{S_{10} + S_{20}} = \frac{\Pi^2 - 1}{\Pi}$$

$$\frac{\sqrt{\Delta^2 + 4(S_{10} + S_{20})^2}}{2(S_{10} + S_{20})} = \frac{\Pi^2 + 1}{2\Pi} \tag{8.45}$$

$$\frac{1}{\sqrt{\Delta^2 + 4(S_{10} + S_{20})^2}} = \frac{1}{S_{10} + S_{20}} \frac{\Pi}{\Pi^2 + 1}$$

Hence differentiating with respect to S_{10} or S_{20} the terms of expressions (8.44) one finds

$$\frac{\partial S_{1i}}{\partial S_{10}} = \sigma_{11} = \frac{1}{\Pi}\left(1 + \frac{S_{10}}{S_{10} + S_{20}} \frac{\Pi^2 - 1}{\Pi^2 + 1}\right)$$

$$\frac{\partial S_{1i}}{\partial S_{20}} = \sigma_{12} = \frac{1}{\Pi} \frac{S_{10}}{S_{10} + S_{20}} \frac{\Pi^2 - 1}{\Pi^2 + 1} \tag{8.46}$$

$$\frac{\partial S_{2i}}{\partial S_{10}} = \sigma_{21} = \frac{1}{\Pi} \frac{S_{20}}{S_{10} + S_{20}} \frac{\Pi^2 - 1}{\Pi^2 + 1}$$

It appears from these relationships that the σ coefficients are not independent and should follow the relationships

$$\sigma_{11} + \sigma_{21} = \sigma_{22} + \sigma_{12} = \frac{2\Pi}{\Pi^2 + 1} \tag{8.47}$$

$$\sigma_{11} + \sigma_{22} - \sigma_{12} - \sigma_{21} = \frac{2}{\Pi}$$

There is an obvious advantage in expressing these variables and parameters in dimensionless form. One has

$$\alpha_0 = \frac{S_{10}}{K_1} \quad \beta_0 = \frac{S_{20}}{K_2}$$

$$\delta = \frac{\Delta}{K_2} \quad \varepsilon = \frac{K_1}{K_2} \quad \lambda = \frac{V_2}{V_1} \quad \theta = \frac{V_1}{K_1}t \tag{8.48}$$

$$v_i = \frac{v_i}{V_1} \quad v_1 = \frac{v_1}{V_1} \quad v_0 = \frac{v_0}{V_1} \quad v_2 = \frac{v_2}{V_1}$$

Moreover one can derive the expression of the partition coefficient Π in terms of the same dimensionless parameters. One finds

$$\Pi = \frac{\sqrt{\delta^2 + 4(\varepsilon\alpha_0 + \beta_0)^2} + \delta}{2(\varepsilon\alpha_0 + \beta_0)} = \frac{2(\varepsilon\alpha_0 + \beta_0)}{\sqrt{\delta^2 + 4(\varepsilon\alpha_0 + \beta_0)^2} - \delta} \tag{8.49}$$

and the sensitivity coefficients can also be written in terms of the same dimensionless variables and parameters. One finds

$$\frac{\partial S_{1i}}{\partial S_{10}} = \sigma_{11} = \frac{1}{\Pi}\left(1 + \frac{\varepsilon\alpha_0}{\varepsilon\alpha_0 + \beta_0}\frac{\Pi^2 - 1}{\Pi^2 + 1}\right)$$

$$\frac{\partial S_{1i}}{\partial S_{20}} = \sigma_{12} = \frac{1}{\Pi}\frac{\varepsilon\alpha_0}{\varepsilon\alpha_0 + \beta_0}\frac{\Pi^2 - 1}{\Pi^2 + 1} \tag{8.50}$$

$$\frac{\partial S_{2i}}{\partial S_{10}} = \sigma_{21} = \frac{1}{\Pi}\frac{\beta_0}{\varepsilon\alpha_0 + \beta_0}\frac{\Pi^2 - 1}{\Pi^2 + 1}$$

$$\frac{\partial S_{2i}}{\partial S_{20}} = \sigma_{22} = \frac{1}{\Pi}\left(1 + \frac{\beta_0}{\varepsilon\alpha_0 + \beta_0}\frac{\Pi^2 - 1}{\Pi^2 + 1}\right)$$

The system (8.41) can now be expressed in matrix form as

$$\frac{d}{d\theta}\begin{bmatrix} \alpha_0 \\ \beta_0 \end{bmatrix} = \Omega^{-1}\begin{bmatrix} 1+\sigma_{22} & -\sigma_{12} \\ -\varepsilon\sigma_{21} & (1+\sigma_{11})\varepsilon \end{bmatrix}\begin{bmatrix} v_i - v_1 + v_2 \\ v_1 - v_2 - v_0 \end{bmatrix} \tag{8.51}$$

where $\Omega = \dfrac{(\Pi+1)^3}{\Pi(\Pi^2+1)}$ \hfill (8.52)

In system (8.51) the normalized rates v_i, v_1, v_2 and v_0 are functions of α_0 and β_0. These functions can be denoted $F_1(\alpha_0, \beta_0)$ and $F_2(\alpha_0, \beta_0)$. Hence the system (8.51) above can be rewritten as

$$\frac{d}{d\theta}\begin{bmatrix} \alpha_0 \\ \beta_0 \end{bmatrix} = \begin{bmatrix} F_1(\alpha_0, \beta_0) \\ F_2(\alpha_0, \beta_0) \end{bmatrix} \tag{8.53}$$

As

$$v_i - v_1 + v_2 = u_1 \tag{8.54}$$

$$v_1 - v_2 - v_0 = u_2$$

one has under steady state

$$\frac{\partial u_1^*}{\partial \alpha_0} = \frac{\partial v_2^*}{\partial \alpha_0} - \frac{\partial v_1^*}{\partial \alpha_0}$$

$$\frac{\partial u_2^*}{\partial \alpha_0} = \frac{\partial v_1^*}{\partial \alpha_0} - \frac{\partial v_2^*}{\partial \alpha_0} \tag{8.55}$$

$$\frac{\partial u_1^*}{\partial \beta_0} = \frac{\partial v_2^*}{\partial \beta_0} - \frac{\partial v_1^*}{\partial \beta_0}$$

$$\frac{\partial u_2^*}{\partial \beta_0} = \frac{\partial v_1^*}{\partial \beta_0} - \frac{\partial v_2^*}{\partial \beta_0} - \mu$$

It follows from these relationships that

$$\frac{\partial u_2^*}{\partial \alpha_0} = -\frac{\partial u_1^*}{\partial \alpha_0} \qquad (8.56)$$

$$\frac{\partial u_2^*}{\partial \beta_0} = -\frac{\partial u_1^*}{\partial \beta_0} - \mu$$

System (8.53) possesses a steady state if

$$\left. \begin{array}{c} \overline{F_1^*(\alpha_0, \beta_0)} \\ \underline{F_2^*(\alpha_0, \beta_0)} \end{array} \right] = 0 \qquad (8.57)$$

In these expressions the starred symbols refer to a steady state. Stability analysis of the system is based on the trace, T_J and on the determinant, Δ_J, of the Jacobian matrix

$$J = \begin{bmatrix} \partial F_1^* / \partial \alpha_0 & \partial F_1^* / \partial \beta_0 \\ \partial F_2^* / \partial \alpha_0 & \partial F_2^* / \partial \beta_0 \end{bmatrix} \qquad (8.58)$$

as well as on its discriminant $T_J^2 - 4\Delta_J$.

The differential system (8.53) is build up on the two functions

$$F_1(\alpha_0, \beta_0) = \Omega^{-1}(1 + \sigma_{22})u_1 - \Omega^{-1}\sigma_{12}u_2 \qquad (8.59)$$

$$F_2(\alpha_0, \beta_0) = -\varepsilon\Omega^{-1}\sigma_{21}u_1 + \varepsilon\Omega^{-1}(1 + \sigma_{11})u_2$$

Under steady state conditions $u_1^* = u_2^* = 0$ and the elements of the jacobian matrix can be expressed to

$$\frac{\partial F_1^*}{\partial \alpha_0} = \Omega^{*-1}\left\{(1 + \sigma_{22}^*)\frac{\partial u_1^*}{\partial \alpha_0} - \sigma_{12}^*\frac{\partial u_2^*}{\partial \alpha_0}\right\}$$

$$\frac{\partial F_1^*}{\partial \beta_0} = \Omega^{*-1}\left\{(1 + \sigma_{22}^*)\frac{\partial u_1^*}{\partial \beta_0} - \sigma_{12}^*\frac{\partial u_2^*}{\partial \beta_0}\right\} \qquad (8.60)$$

$$\frac{\partial F_2^*}{\partial \alpha_0} = -\varepsilon \Omega^{*-1} \left\{ \sigma_{21}^* \frac{\partial u_1^*}{\partial \alpha_0} - (1+\sigma_{11}^*) \frac{\partial u_2^*}{\partial \alpha_0} \right\}$$

$$\frac{\partial F_2^*}{\partial \beta_0} = -\varepsilon \Omega^{*-1} \left\{ \sigma_{21}^* \frac{\partial u_1^*}{\partial \beta_0} - (1+\sigma_{11}^*) \frac{\partial u_2^*}{\partial \beta_0} \right\}$$

Taking advantage of expressions (8.56) these equations can be simplified. One finds

$$\frac{\partial F_1^*}{\partial \alpha_0} = \Omega^{*-1} \left\{ 1+\sigma_{12}^* + \sigma_{22}^* \right\} \frac{\partial u_1^*}{\partial \alpha_0}$$

$$\frac{\partial F_2^*}{\partial \alpha_0} = -\varepsilon \Omega^{*-1} \left\{ 1+\sigma_{11}^* + \sigma_{21}^* \right\} \frac{\partial u_1^*}{\partial \alpha_0} \tag{8.61}$$

$$\frac{\partial F_1^*}{\partial \beta_0} = \Omega^{*-1} \left\{ (1+\sigma_{22}^* + \sigma_{12}^*) \frac{\partial u_1^*}{\partial \beta_0} + \mu \sigma_{12}^* \right\}$$

$$\frac{\partial F_2^*}{\partial \beta_0} = -\varepsilon \Omega^{*-1} \left\{ (1+\sigma_{11}^* + \sigma_{21}^*) \frac{\partial u_1^*}{\partial \beta_0} + (1+\sigma_{11}^*)\mu \right\}$$

One can express Ω^{*-1} and the σ coefficients as a function of the steady state partition coefficient Π^*. Setting for simplicity

$$\rho^* = \frac{\varepsilon \alpha_0^*}{\varepsilon \alpha_0^* + \beta_0^*} \tag{8.62}$$

one has

$$\Omega^{*-1}(1+\sigma_{11}^* + \sigma_{21}^*) = \Omega^{*-1}(1+\sigma_{12}^* + \sigma_{22}^*) = \frac{\Pi^*}{1+\Pi^*}$$

$$\mu \Omega^{*-1} \sigma_{12}^* = \rho^* \frac{\Pi^* - 1}{(\Pi^* + 1)^2} \tag{8.63}$$

$$\mu\Omega^{*-1}(1+\sigma_{11}^*) = \frac{\Pi^{*2}+1}{(\Pi^*+1)^2}\mu + \rho^* \frac{\Pi^*-1}{(\Pi^*+1)^2}\mu$$

and it becomes possible to express the terms of the jacobian matrix in terms of the electrostatic partition coefficient. One has

$$\frac{\partial F_1^*}{\partial\alpha_0} = \frac{\Pi^*}{\Pi^*+1}\frac{\partial u_1^*}{\partial\alpha_0}$$

$$\frac{\partial F_2^*}{\partial\alpha_0} = -\varepsilon\frac{\Pi^*}{\Pi^*+1}\frac{\partial u_1^*}{\partial\alpha_0}$$ (8.64)

$$\frac{\partial F_1^*}{\partial\beta_0} = \frac{\Pi^*}{\Pi^*+1}\frac{\partial u_1^*}{\partial\beta_0} + \rho^*\frac{\Pi^*-1}{(\Pi^*+1)^2}\mu$$

$$\frac{\partial F_2^*}{\partial\beta_0} = -\varepsilon\frac{\Pi^*}{\Pi^*+1}\frac{\partial u_1^*}{\partial\beta_0} - \varepsilon\frac{\Pi^{*2}+1}{(\Pi^*+1)^2}\mu - \varepsilon\rho^*\frac{\Pi^*-1}{(\Pi^*+1)^2}\mu$$

If we express the trace, T_J, and the determinant of the jacobian matrix in terms of the derivatives of the reaction rates v_1^* and v_2^* with respect of α_0 and β_0, one finds

$$T_J = \frac{\Pi^*}{\Pi^*+1}\left\{\left(\frac{\partial v_2^*}{\partial\alpha_0} - \frac{\partial v_1^*}{\partial\alpha_0}\right) + \varepsilon\left(\frac{\partial v_1^*}{\partial\beta_0} - \frac{\partial v_2^*}{\partial\beta_0}\right)\right\} - \varepsilon\frac{\Pi^{*2}+1}{(\Pi^*+1)^2}\mu - \varepsilon\rho^*\frac{\Pi^*-1}{(\Pi^*+1)^2}\mu \quad (8.65)$$

and

$$\Delta_J = \varepsilon\frac{\Pi^*(\Pi^{*2}+1)}{(\Pi^*+1)^3}\left(\frac{\partial v_1^*}{\partial\alpha_0} - \frac{\partial v_2^*}{\partial\alpha_0}\right)\mu$$ (8.66)

One can notice that if there is no attraction, or repulsion, effects exerted on the reagents, then the electrostatic partition coefficient is equal to one, v_1 is a function of α_0 only and v_2 a function of β_0 only. If the two enzymes follow Michaelis-Menten kinetics $\partial v_2^*/\partial\alpha_0 = \partial v_1^*/\partial\beta_0$ and one has

$$T_J = -\frac{1}{2}\left(\frac{\partial v_1^*}{\partial \alpha_0} + \varepsilon \frac{\partial v_2^*}{\partial \beta_0}\right) - \frac{1}{2}\varepsilon\mu \tag{8.67}$$

$$\Delta_J = \frac{1}{4}\varepsilon\mu\frac{\partial v_1^*}{\partial \alpha_0}$$

Under these conditions $T_J < 0, \Delta_J > 0$. Then the system can only display a stable node.

A condition required to generate "atypic" behavior of the system can be observed if the charged medium exert electrostatic repulsion effects on substrate β. One can notice that if there is no attraction, or repulsion, effects exerted on the reagents, then the electrostatic partition coefficient is equal to one. Under these conditions, v_1 is a function of α_0 only and v_2 a function of β_0 only. If the two enzymes follow Michaelis-Menten kinetics then $\partial v_2^* / \partial \alpha_0 = 0$ and $\partial v_1^* / \partial \beta_0 = 0$. Under these conditions one has

$$T_J = -\frac{1}{2}\left(\frac{\partial v_1^*}{\partial \alpha_0} + \varepsilon \frac{\partial v_2^*}{\partial \beta_0}\right) - \frac{1}{2}\varepsilon\mu \tag{8.68}$$

$$\Delta_J = \frac{1}{4}\varepsilon\mu\frac{\partial v_1^*}{\partial \alpha_0}$$

and the system can only display a stable node for $T_J < 0, \Delta_J > 0$ and $T_J^2 - 4\Delta_J > 0$. A condition required to generate atypical behavior of the system can be observed if the charged medium exerts electrostatic repulsion on substrate β. In this case, the rate v_2 becomes a function of both α_0 and β_0. Let us consider the two reaction rates

$$v_1 = \frac{\alpha_0}{1 + \alpha_0} \tag{8.69}$$

$$v_2 = \frac{\lambda\beta_0}{\Pi + \beta_0}$$

the second equation displays attraction-repulsion effects of α_0 and β_0 by the charged matrix. One has

$$\frac{\partial v_1^*}{\partial \alpha_0} = \frac{1}{(1+\alpha_0)^2}$$

$$\frac{\partial v_2^*}{\partial \alpha_0} = -\lambda \frac{\beta_0(\partial \Pi^* / \partial \alpha_0)}{(\Pi^* + \beta_0)^2} \qquad (8.70)$$

$$\frac{\partial v_2^*}{\partial \beta_0} = \lambda \frac{\Pi^* - \beta_0(\partial \Pi^* / \partial \beta_0)}{(\Pi^* + \beta_0)^2}$$

It is worth stressing that, contrary to v_1^*, the rate v_2^* depends upon both α_0 and β_0. As $\partial v_1^* / \partial \beta_0 = 0$, the trace and the determinant of the jacobian matrix assume the form

$$T_J = \frac{\Pi^*}{\Pi^* + 1} \left\{ \frac{1}{(1+\alpha_0)^2} + \varepsilon \frac{\lambda \Pi^*}{(\Pi^* + \beta_0)^2} \right\} - \varepsilon \frac{\Pi^* + 1}{(\Pi^* + 1)^2} \mu - \varepsilon \rho^* \frac{\Pi^* - 1}{(\Pi^* + 1)^2} \mu \qquad (8.71)$$

$$\Delta_J = \varepsilon \frac{\Pi^*(\Pi^{*2} + 1)}{(\Pi^* + 1)^3} \left\{ \frac{1}{(1+\alpha_0)^2} + \frac{\lambda \beta_0}{(\Pi^* + \beta_0)^2} \frac{\partial \Pi^*}{\partial \alpha_0} \right\}$$

As $\partial \Pi^* / \partial \alpha_0$ is negative, the expression for Δ_J can be negative as well. This situation generates a complete change of the dynamics of the system that displays a saddle point.

Electric repulsion effects can generate multistability of the system (8.69). Under these conditions one has

$$Y_1(\alpha_0) = v_1(\alpha_0) = \frac{\alpha_0}{1+\alpha_0} \qquad (8.72)$$

$$Y_2(\alpha_0) = \mu\beta* + \frac{\lambda\beta_0^*}{\Pi + \beta_0^*}$$

It then appears that Π^* is a decreasing function of α_0. As a matter of fact, one has

$$\frac{\partial \Pi^*}{\partial \alpha_0} = \frac{\varepsilon\delta\left\{\sqrt{\left[\delta^2 + 4(\varepsilon\alpha_0 + \beta_0)^2\right]} + \delta\right\}}{2(\varepsilon\alpha_0 + \beta_0)^2 \sqrt{\left[\delta^2 + 4(\varepsilon\alpha_0 + \beta_0)^2\right]}} \tag{8.73}$$

Moreover if follows from the second equation (8.72) that

$$\frac{\partial Y_2(\alpha_0)}{\partial \alpha_0} = -\frac{\lambda\beta_0^*}{(\Pi + \beta_0^*)^2}\frac{\partial \Pi}{\partial \alpha_0} \tag{8.74}$$

and it appears that $Y_2(\alpha_0)$ is an increasing function of α_0. If the curves of equations $Y_1(\alpha_0)$ and $Y_2(\alpha_0)$ are plotted as a function of α_0 the system can have two steady states of which one is stable and the other unstable (Fig. **6**). Moreover if the partition coefficient Π is plotted as a function of β_0 it appears that it possesses at most three different values (Fig. **7**).

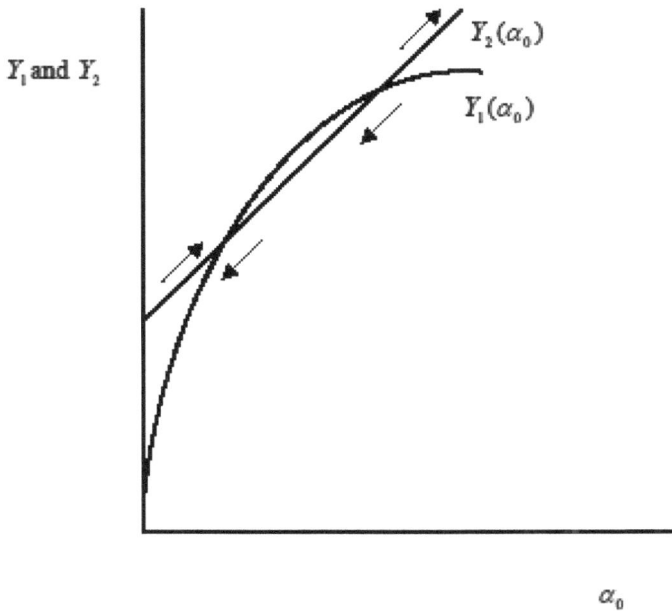

Figure 6: Multiple steady states for the open metabolic cycle. Two steady states, one stable and the other one unstable, are obtained for $Y_1(\alpha_0)$ and $Y_2(\alpha_0)$.

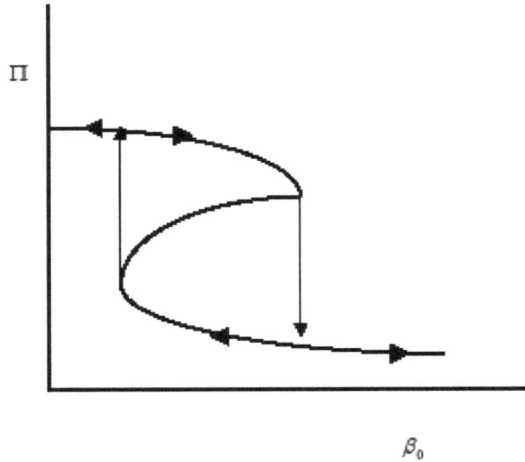

Figure 7: Multiple steady state of the partition coefficient Π. See text.

When the bulk concentration of a charged substrate is varied the electrostatic partition coefficient Π varies as well. This implies that the local pH changes. More specifically, the local pH increases as the substrate concentration is increased. We postulate, for simplicity, that one substrate, S_1, is neutral whereas the other one, S_2, is a monoanion. Moreover the normalized bulk proton concentration γ_b is defined as $\gamma_b = H_0 / K_b$ where K_b is the base ionization constant of the enzyme-substrate complex. If the volume of the bulk phase is much larger than that of the matrix, γ_b is roughly constant. If, in addition to that, we assume that the reaction is sensitive to high pH values then

$$\tilde{V}_2 = V_2 \frac{H_i}{K_b + H_i} = V_2 \frac{H_0 \Pi}{K_B + H_0 \Pi} = V_2 \frac{\gamma_b \Pi}{1 + \gamma_B \Pi} \tag{8.75}$$

where $\tilde{V}_2$ and V_2 are the apparent and the real maximum reaction rate, respectively. The two rate equations are then

$$v_1(\alpha_0) = \frac{\alpha_0}{1 + \alpha_0} \tag{8.76}$$

$$v_2(\beta_0) = \lambda \frac{\gamma_b \Pi}{1 + \gamma_b \Pi} \frac{\beta_0}{\Pi + \beta_0}$$

If $\gamma_b > 1$, the function $v_2(\beta_0)$ is monotonic and increasing, but if $\gamma_b < 1$ this function reaches a maximum and then decreases as β_0 increases. It then appears that although the enzyme follows *per se* Michaelis-Menten kinetics it displays apparent inhibition by excess substrate. This situation is depicted in Fig. **8**.

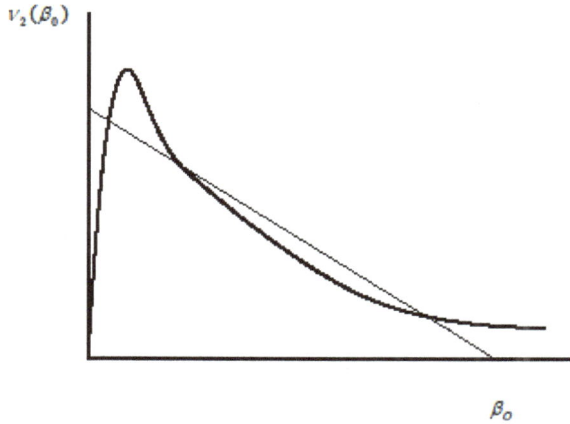

Figure 8: Multiple steady states of the function $v_2(\beta_0)$. The three steady states pertain to the intersections of the functions $v_2(\beta_0)$ and $\mu(\beta_0)$.

If, as previously assumed, S_1 is an uncharged molecule and S_2 a monoanion, one has

$$\sigma_{11} = 1$$

$$\sigma_{12} = 0 \tag{8.77}$$

$$\sigma_{21} = 0$$

$$\sigma_{22} = \frac{2\Pi}{\Pi^2 + 1}$$

and the expression of the corresponding dynamic system is then

$$\frac{d}{d\theta}\begin{bmatrix} \alpha_0 \\ \beta_0 \end{bmatrix} = \Omega^{-1}\begin{bmatrix} 1+\sigma_{22} & 0 \\ 0 & 2\varepsilon \end{bmatrix}\begin{bmatrix} u_1 \\ u_2 \end{bmatrix} \tag{8.78}$$

with

$$\Omega = \frac{2(\Pi+1)^2}{\Pi^2+1} \tag{8.79}$$

and

$$u_1 = v_i - v_1(\alpha_0) + v_2(\beta_0) \tag{8.80}$$

$$u_2 = v_1(\alpha_0) - v_2(\beta_0) - \mu\beta_0$$

The steady state values of the state variables α and β should be a solution of

$$v_i - v_1(\alpha_0) + v_2(\beta_0^*) = 0 \tag{8.81}$$

$$v_1(\alpha_0) - v_2(\beta_0^*) - \mu\beta_0^* = 0$$

which implies that

$$\beta_0^* = \frac{v_i}{\mu} \tag{8.82}$$

If we consider the system (8.78) it can be rewritten as

$$F_1(\alpha_0,\beta_0) = \frac{1}{2}\{v_i - v_1(\alpha_0) + v_2(\beta_0)\} \tag{8.83}$$

$$F_2(\alpha_0,\beta_0) = \varepsilon \frac{\Pi^2+1}{(\Pi+1)^2}\{v_1(\alpha_0) - v_2(\beta_0) - \mu\beta_0\}$$

The jacobian matrix of the system is

$$J = \begin{bmatrix} \partial F_1^* / \partial\alpha_0 & \partial F_1^* / \partial\beta_0 \\ \partial F_2^* / \partial\alpha_0 & \partial F_2^* / \partial\beta_0 \end{bmatrix} \tag{8.84}$$

and taking account of expressions (8.83) one has

$$\frac{\partial F_1^*}{\partial \alpha_0} = -\frac{1}{2}\frac{\partial v_1^*}{\partial \alpha_0} \qquad\qquad \frac{\partial F_1^*}{\partial \beta_0} = \frac{1}{2}\frac{\partial v_1^*}{\partial \beta_0} \qquad\qquad (8.85)$$

$$\frac{\partial F_2^*}{\partial \alpha_0} = \varepsilon\frac{\Pi^{*2}+1}{(\Pi^*+1)^2}\frac{\partial v_1^*}{\partial \alpha_0} \qquad \frac{\partial F_2^*}{\partial \beta_0} = -\varepsilon\frac{\Pi^{*2}+1}{(\Pi^*+1)^2}\left(\frac{\partial v^2}{\partial \beta_0}+\mu\right)$$

The trace, T_J, and the determinant, Δ_J, of the jacobian matrix (8.84) are then

$$T_J = -\left\{\frac{1}{2}\frac{\partial v_1^*}{\partial \alpha_0} + \varepsilon\frac{\Pi^{*2}+1}{(\Pi^*+1)^2}\left(\frac{\partial v_2^*}{\partial \beta_0}+\mu\right)\right\} \qquad (8.86)$$

$$\Delta_J = \frac{\varepsilon}{2}\frac{\Pi^{*2}+1}{(\Pi^*+1)^2}\mu\frac{\partial v_1^*}{\partial \alpha_0}$$

From the trace and the determinant of the jacobian matrix one can derive the expression of the corresponding discriminant, $T_J^2 - 4\Delta_J$, of the matrix. One finds

$$T_J^2 - 4\Delta_J = \left\{\frac{1}{2}\frac{\partial v_1^*}{\partial \alpha_0} + \varepsilon\frac{\Pi^{*2}+1}{(\Pi^*+1)^2}\left(\frac{\partial v_2^*}{\partial \beta_0}+\mu\right)\right\}^2 - 2\varepsilon\frac{\Pi^{*2}+1}{(\Pi^*+1)^2}\mu\frac{\partial v_1^*}{\partial \alpha_0} \qquad (8.87)$$

Hence it appears that a necessary, but not sufficient, condition to obtain sustained oscillations of α_0 and β_0 is $\partial v_2^*/\partial \beta_0 < 0$. If this condition is fulfilled the trace T_j and the discriminant $T_J^2 - 4\Delta_J$ are both negative. Moreover as β_0 can oscillate the electric partition coefficient can oscillate as well (Fig. **9**).

4- GENERAL CONCLUSIONS

There are several examples of oscillatory enzyme reactions that take place in living cells. Probably the most studied so far are periodic and aperiodic oscillations of glycolysis [4]. The term "oscillatory enzyme" can sometimes be found in scientific literature, even though there is no evidence that such an enzyme exists. So far no enzyme has been isolated that displays periodic behaviour. The minimum system able to exhibit such a behavior is made up of two antagonistic enzyme reactions displaying non-linear terms.

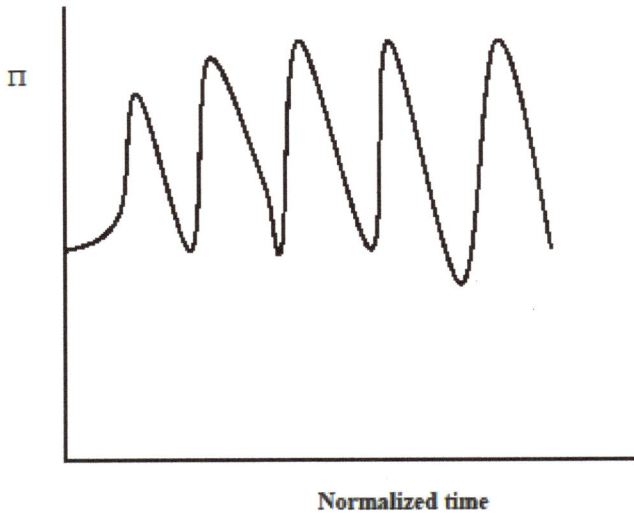

Figure 9: Periodic variations of the electrostatic partition coefficient Π.

In fact it is highly probable that many enzyme reactions take place *in vivo* on membranes or cell walls. Both membranes and cell walls are bearing fixed negative charges that may interact with mobile charges of the substrates of the enzyme reaction. Owing to these electrostatic interactions one can observe different effects that could play a biological function. For instance, an enzyme reaction associated with an insoluble polyelectrolyte (a membrane or a cell wall) can display hysteretic effects *i.e.* the enzyme reaction rate will be different, for the same substrate concentration, depending on this concentration is actually reached through an increase or a decrease. The attraction-repulsion effects superimposed to the standard catalytic process may also generate regular sustained, or chaotic, oscillations of the enzyme reaction rate. These effects are probably not, in most cases, a property of the enzyme but of a global system that includes the enzyme reaction.

REFERENCES

[1]   Goldbeter, A. (1996) Biochemical Oscillations and Biological Rythms. Cambridge University Press, Cambridge.
[2]   Goldbeter, A. and Caplan, S.R. (1976) Oscillatory Enzymes. Annu; Rev. Biophys. Bioeng. 5, 369-476.
[3]   Goldbeter, A., Dupont, G. and Berridge, M. (1990) Minimal model for a signal–induced calcium oscillations and for their frequency encoding through protein phosphorylation. Proc. Natl. Acad. Sci. USA 87, 1461-1465.

[4]   Hess, B. and Boiteux, A. (1971) Oscillatory phenomena in biochemistry. Annu. Rev. Biochem. 40, 237-258.

[5]   Hess, B. (1997) Periodical patterns in biochemical reactions. Quart. Rev. Biophys. 30, 121-176.

[6]   Markus, M. and Hess, B.(1984) Transitions between oscillatory modes in a glycolytic model system. Proc. Natl. Acad. USA 81, 4394-4398.

[7]   Jacob, R., Meritt, J. E., Hallam, T. J. and Rink, T. J. (1988) Repetitive spikes in cytoplasmic calcium evoked by histamine in human endothelial cells. Nature, 335, 40-45.

[8]   Meyer, T. and Stryer, L.(1991) Calcium spiking. Annu. Rev. Biophys. Biophys. Chem. 20, 153-174.

[9]   Nicolis, G. and Prigogine, I. (1977) Self-organization in Nonequilibrium Systems. From Dissipative Structures to Order through Fluctuations. John Wiley and Sons. New York.

[10]   Guidi, G. M., Carlier, M. F. and Goldbeter, A. (1998) Bistability in the isocitrate dehydrogenase reaction: an experimentally based theoretical study. Biophys. J. 74, 1229-1240.

[11]   Ricard, J. (1989) Modulation of Enzyme Catalysis in Organized Biological Systems. Catalysis Today. 5, 275-384.

[12]   Ricard, J., Mulliert, G., Kellershohn, N. and Giudici-Orticoni, M. T. (1994) Dynamics of enzyme reactions and metabolic networks in living cells. A physico-chemical approach. Prog. Mol. Subcell. Biol. 13, 1-80.

[13]   Kellershohn, N., Mulliert, G. and Ricard, J. (1990) Dynamics of an open futile cycle at the surface of a charged membrane. I. A simple general model. Physica D 46, 367-374.

[14]   Mulliert, G., Kellershohn, N. and Ricard, J. (1990) Dynamics of an open futile cycle at the surface of a charged membrane. II Multiple steady states and oscillatory behavior generated by electric repulsion effects. Physica D 46, 380-391.

[15]   Ricard, J., Kellershohn, N. and Mulliert, G. (1992) Dynamic aspects of long distance functional interactions between membrane-bound enzymes. J. Theor. Biol. 156, 1640.

[16]   Ricard, J. Biological Complexity and the Dynamics of Life Processes.(1999) Elsevier, Amsterdam, Lausanne, New York.

Send Orders for Reprints to reprints@benthamscience.net

CHAPTER 9

Biochemical Networks

Abstract: Any set of connected enzyme reactions constitute a network. Contrary to most networks studied so far (for instance networks of social connections) metabolic networks are open structures with an input and an output of matter. As such they possess some properties that are unique when compared to closed networks.

Keywords: Node degree, Random graph, Diameter of a graph, Closed network, Open network, Meta-network, Conditional probabilities of binding for a network, Information consumed, or produced, by a node, Open character of a network and probabilities of occurrence of its nodes, Thermodynamic constraints upon running from a node to another one.

Material elements that interact form a network that can be represented, from a mathematical viewpoint, by a graph also called network. The nodes of the graph are the mathematical description of a population of elements in interaction, and the edges represent the interactions that exist between the nodes. Rather recently the study of networks has undergone a spectacular development. As networks are present nearly everywhere in the world the concept of graph has transgressed the boundaries between different scientific fields. In this line, nearly every kind of collective property can be described by a network. In biology, networks are present at all levels, molecular (enzymes and metabolic networks), cellular (genetic networks), multicellular (neural networks), organismic (predator-prey networks) and population (social networks). The science of networks has expanded very rapidly crossing the boundaries between different disciplines in such a way it could be considered the science of connections between material entities whatever their nature....

1- THE CONCEPT OF NETWORK

A network can be represented, regardless of its nature, by a graph. As mentioned previously, a graph is a set of points, the nodes, connected by the edges according to a certain topology. From a mathematical viewpoint, a graph G is a subset of the cartesian product of the set of nodes S by itself. According to this definition one has

Jacques Ricard

$$G \subset S \times S \tag{9.1}$$

If x_i and y_j ($\forall x_i, y_j \in S$) are the nodes of the graph we associate x_i and y_j according to a certain relation $x_i R y_j$. If, for instance, the set is $S = \{1, 2, 3, 4, 5\}$ and the relation R means $x_i < y_j$ the graph is shown in the Fig. **1** and may be represented by the following binary matrix

$$
x_i \downarrow \quad
\begin{array}{c c}
 & \overline{\qquad} \; y_J \longrightarrow \\
\begin{array}{c} 1 \\ 2 \\ 3 \\ 4 \\ 5 \end{array} &
\begin{array}{ccccc}
1 & 2 & 3 & 4 & 5 \\
\left[\begin{array}{ccccc}
0 & 1 & 1 & 1 & 1 \\
0 & 0 & 1 & 1 & 1 \\
0 & 0 & 0 & 1 & 1 \\
0 & 0 & 0 & 0 & 1 \\
0 & 0 & 0 & 0 & 0
\end{array}\right]
\end{array}
\end{array}
$$

This matrix describes the graph shown in Fig. **1**.

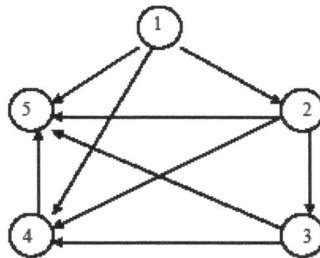

Figure 1: The graph pertaining to the above matrix. See text.

It has often been considered that large networks are random. Such networks have been studied by Erdos and Renyi [1] as well as by Bolobas [2]. Given a fixed number of nodes one can build up differents graphs. Moreover it is intuitively evident that increasing the number of connections leads to a decrease of the number of possible graphs and finally one obtains a unique but giant network. The connection of a node to others is called a node degree d. One can therefore define, for any random graph, a mean node degree $<k>$. The diameter $<d>$ of the graph is defined as

$$<d> = \frac{\ln N}{\ln <k>} \tag{9.2}$$

where N is the number of nodes. Moreover in such random systems the node degree should be distributed according to a Poisson law [2]. This implies in practice that few nodes are highly, or poorly, connected.

Sequencing the genome of different living organisms has allowed to build up the network displaying the relationships between all the proteins present in living organisms [3-5]. It is then possible to build up metabolic networks spanning the phylogenetic tree from mycoplasms and bacteria up to eukaryotes [6]. In these networks, each node is a metabolite and the edge connecting two nodes is an enzyme reaction allowing the conversion of a metabolite into another one. What has been observed is that the diameter of the network remains unchanged upon increasing the number of nodes. Moreover, in such networks, if one plots the average number of incoming and outcoming links as a function of the number of nodes one obtains an increasing curve. It therefore appears that node connectivity increases with the number of nodes [6]. According to Barabasi and his colleagues there should exist a mechanism that would keep the network diameter constant whatever the number of metabolites. Moreover, in this perspective, there should exist an evolutionary pressure that tends to generate an irregular degree of connection of the nodes. Such networks have been called small-world for they should possess short paths between distant nodes. On can speculate, in a neo-Darwinian perspective, about the possible advantages such a situation could offer. There are, at least, two functional advantages. The first one implies that whatever the complexity of the metabolic networks there are always some paths that allow rapid conversion of metabolites. Moreover removing nodes in a random fashion does not change dramatically the diameter of small-world networks. Moreover if metabolic networks are constructed from data bases pertaining to organisms spanning the phylogenetic tree from mycoplasms and bacteria to eukaryotes the network diameter does not vary significantly. This result has been interpreted as a kind of strategy that allows these organisms to resist random mutations of their genomes [6].

2- METABOLIC NETWORKS AS "OPEN" SYSTEMS

If we consider a metabolic network, it can be viewed as being up made up of connected enzyme reactions. This means that each node, called macro-node, is *per se*, an enzyme reaction. If we assume that every enzyme reaction is fast relative to the rate of transport of intermediates from enzyme to enzyme within the network, every node (enzyme reaction) can be considered as an entity. One can immediately realize that this mode of description of biochemical networks is completely different from the classical one used by Barabasi and co-workers [7-10]. In the approach presented in this book the vision of metabolic networks is dynamic and matches reality for two reasons:

- The first is to consider a metabolic network as a network of enzyme reactions and not as a network of connected chemicals;

- The second one is that this mode of description of metabolic networks puts the emphasis on the fact that such systems are open, that is they receive an input, and release an output, of matter that will be used in another network. In this perspective, a metabolic system is an open system of connected enzyme reactions and not a *closed* system of connected chemicals.

It then appears that any macro-node is an enzyme reaction connected to its neighbours. Let us consider a macro-node of a metabolic network, the probability that enzyme E_i has bound substrate A_i is

$$p(A_i)_{Ei} = \frac{[E_i A_i] + [E_i A_i B_i]}{Y_i} \qquad (9.3)$$

where Y_i is the expression of the total concentration of enzyme E_i that binds substrates A_i and B_i. One has then

$$Y_i = [E_i] + [E_i A_i] + [E_i A_i B_i] \qquad (9.4)$$

It follows from this definition that the probability that the metabolic network, defined as a meta-network N, has bound substrate A_i is

$$p(A_i)_N = \frac{[E_i A_i] + [E_i A_i B_i]}{Y_T} \qquad (9.5)$$

where

$$Y_T = \sum_{i=1}^{n} Y_i \qquad (9.6)$$

As

$$p(Y_i) = \frac{Y_i}{Y_T} \tag{9.7}$$

one has

$$p(A_i)_N = \frac{[E_iA_i]+[E_iA_iB_i]}{Y_i}\frac{Y_i}{Y_T} = p(A_i)_N\, p(Y_i) \tag{9.8}$$

where $p(Y_i)$ is the probability of occurrence of node Y_i in the meta-network.

Now let us consider the conditional probability that enzyme E_i has bound substrate B_i. The conditional probability $p(A_i|B_i)_{Ei}$ is equal to the conditional probability $p(A_i|B_i)_N$ that the meta-network has bound A_i given it has already bound B_i. Hence one has

$$p(A_i|B_i)_{Ei} = p(A_i|B_i)_N \tag{9.9}$$

It is then possible to define the two functions, $h(A_i)_N$ and $h(A_i|B_i)_N$ as

$$h(A_i)_N = -\log[\,p(y_i)p(A_i)_{Ei}\,] \tag{9.10}$$

and

$$h(A_i|B_i)_N = -\log p(A_i|B_i)_N = -\log p(A_i|B_i)_{Ei} \tag{9.11}$$

It appears that the amount of information consumed or produced at the level of node Y_i of the meta-network is

$$I(A_i:B_i)_N = h(A_i)_N - h(A_i|B_i)_N \tag{9.12}$$

that can be rewritten as

$$I(A_i:B_i)_N = -\log p(y_i) - \log p(A_i)_{Ei} - h(A_i|B_i)_{Ei} \tag{9.13}$$

or as

$$I(A_i : B_i)_N = h(A_i)_{Ei} - h(A_i | B_i)_{Ei} - \log p(Y_i) \tag{9.14}$$

and finally as

$$I(A_i : B_i)_N = I(A_i : B_i)_{Ei} - \log p(Y_i) \tag{9.15}$$

The consequence of this reasoning is that the information produced, or consumed, per node of the meta-network is equal to the information of the same node considered in isolation affected by a term expressing the probability of occurrence of this node. The smaller this probability of occurrence and the larger is the importance of this term. If the number of nodes is large the probability of occurrence of node $Y_i, p(Y_i)$, is very small hence $-\log p(Y_i)$ is very large and positive. It follows from this that the very fact that an enzyme is part of a network gives this enzyme additional information that depends upon network topology. One can express the mean information per node of such a network as

$$< I(A : B) >= \sum_i p(Y_i) I(A_i : B_i)_{Ei} - \sum_i p(Y_i) \log p(Y_i) \tag{9.16}$$

The first term of the right-hand side member of this expression represents the mean contribution of the micro-states. The second term, which can be considered a topological information, expresses how the connexions of the macro-states contribute to the information of the whole system.

3- BIOCHEMICAL NETWORKS AS OPEN SYSTEMS AND THEIR INFORMATION

We have already outlined that biochemical networks are open systems and one may wonder whether this open character plays a part in their informational content. Let us consider an ideal open network made up of four connected enzyme reactions (Fig. **2**). The transition constants, τ_i, from node to node, *i.e.* the transitions constants between enzyme reactions, are in fact the products of different contributions: $p(A_i, B_i)$ the probability of occurrence of the ternary complexes, k_i the corresponding catalytic constants and k_i^{D*} the apparent diffusion constant of this reaction intermediate from enzyme to enzyme. One has

$$\tau_i = k_i k_i^{D*} p(A_i, B_i) \tag{9.17}$$

with

$$k_i^{D*} = k_i^D \frac{\partial [S_i]}{\partial x} \tag{9.18}$$

where k_i^D is the true diffusion constant of substrate S_i. The simple model of multienzyme network is shown in Fig. 1. The system is open with an input (v_i) and an output (τ_0) of matter.

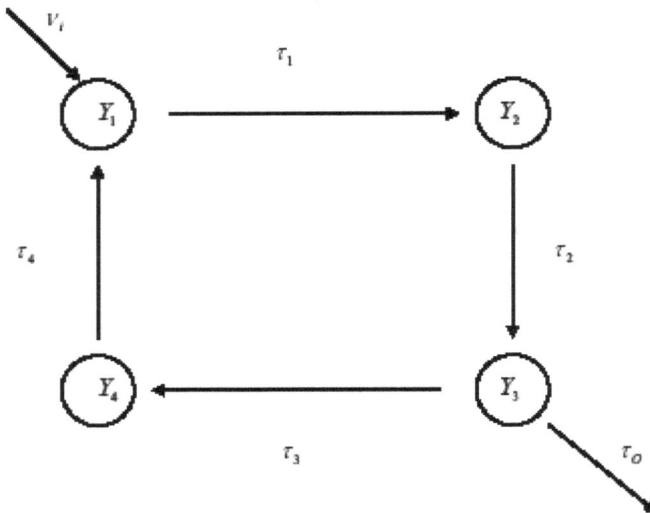

Figure 2: The graph description of an open multienzyme network. Each node in the system is an enzyme reaction.

One can write

$$\frac{dY_1}{dt} = v_i + \tau_4 Y_4 - \tau_1 Y_1$$

$$\frac{dY_2}{dt} = \tau_1 Y_1 - \tau_2 Y_2 \tag{9.19}$$

$$\frac{dY_3}{dt} = \tau_2 Y_2 - (\tau_3 + \tau_0) Y_3$$

$$\frac{dY_4}{dt} = \tau_3 Y_3 - \tau_4 Y_4$$

If the system is in steady state and if Y_T is defined as

$$Y_T = Y_1 + Y_2 + Y_3 + Y_4 \qquad (9.20)$$

the system (9.19) becomes

$$\frac{dp(Y_1)}{dt} = \frac{v_i}{Y_T} + \tau_4 p(Y_4) - \tau_1 p(Y_1) = 0$$

$$\frac{dp(Y_2)}{dt} = \tau_1 p(Y_1) - \tau_2 p(Y_2) = 0 \qquad (9.21)$$

$$\frac{dp(Y_3)}{dt} = \tau_2 p(Y_2) - (\tau_3 + \tau_0) p(Y_3) = 0$$

$$\frac{dp(Y_4)}{dt} = \tau_3 p(Y_3) - \tau_4 p(Y_4) = 0$$

This system can be rewritten in matrix form and one finds

$$\begin{bmatrix} \tau_1 & -\tau_2 & 0 & 0 \\ 0 & \tau_2 & -(\tau_0 + \tau_3) & 0 \\ 0 & 0 & \tau_3 & -\tau_4 \\ 1 & 1 & 1 & 1 \end{bmatrix} \begin{bmatrix} p(Y_1) \\ p(Y_2) \\ p(Y_3) \\ p(Y_4) \end{bmatrix} = \begin{bmatrix} 0 \\ 0 \\ 0 \\ 1 \end{bmatrix} \qquad (9.22)$$

Solving this system yields

$$p(Y_1) = \frac{\tau_2 \tau_3 \tau_4 + \tau_0 \tau_2 \tau_4}{\tau_1 \tau_2 \tau_3 + \tau_1 \tau_2 \tau_4 + \tau_1 \tau_3 \tau_4 + \tau_0 \tau_2 \tau_4 + \tau_0 \tau_1 \tau_4 + \tau_2 \tau_3 \tau_4}$$

$$p(Y_2) = \frac{\tau_1 \tau_3 \tau_4 + \tau_0 \tau_1 \tau_4}{\tau_1 \tau_2 \tau_3 + \tau_1 \tau_2 \tau_4 + \tau_1 \tau_3 \tau_4 + \tau_0 \tau_2 \tau_4 + \tau_0 \tau_1 \tau_4 + \tau_2 \tau_3 \tau_4}$$ (9.23)

$$p(Y_3) = \frac{\tau_1 \tau_2 \tau_4}{\tau_1 \tau_2 \tau_3 + \tau_1 \tau_2 \tau_4 + \tau_1 \tau_3 \tau_4 + \tau_0 \tau_2 \tau_4 + \tau_0 \tau_1 \tau_4 + \tau_2 \tau_3 \tau_4}$$

$$p(Y_4) = \frac{\tau_1 \tau_2 \tau_3}{\tau_1 \tau_2 \tau_3 + \tau_1 \tau_2 \tau_4 + \tau_1 \tau_3 \tau_4 + \tau_0 \tau_2 \tau_4 + \tau_0 \tau_1 \tau_4 + \tau_2 \tau_3 \tau_4}$$

If now one assumes that the system is closed, *i.e.* there is no input and output of matter, the probabilities of Y_1, Y_2, Y_3 and Y_4 are

$$p(Y_1) = \frac{\tau_2 \tau_3 \tau_4}{\tau_1 \tau_2 \tau_3 + \tau_1 \tau_2 \tau_4 + \tau_1 \tau_3 \tau_4 + \tau_2 \tau_3 \tau_4}$$

$$p(Y_2) = \frac{\tau_1 \tau_3 \tau_4}{\tau_1 \tau_2 \tau_3 + \tau_1 \tau_2 \tau_4 + \tau_1 \tau_3 \tau_4 + \tau_2 \tau_3 \tau_4}$$ (9.24)

$$p(Y_3) = \frac{\tau_1 \tau_2 \tau_4}{\tau_1 \tau_2 \tau_3 + \tau_1 \tau_2 \tau_4 + \tau_1 \tau_3 \tau_4 + \tau_2 \tau_3 \tau_4}$$

$$p(Y_4) = \frac{\tau_1 \tau_2 \tau_3}{\tau_1 \tau_2 \tau_3 + \tau_1 \tau_2 \tau_4 + \tau_1 \tau_3 \tau_4 + \tau_2 \tau_3 \tau_4}$$

One may wonder whether the input and output of matter, in and out the system, may alter the probabilities of occurrence of the nodes. In fact two nodes, Y_3 and Y_4, have the same numerator for open an close systems but a different denominator. Two terms are missing in the denominator of Y_3 and Y_4 for a closed system. It is then evident that $p(Y_3)$ and $p(Y_4)$ will be larger in the case of a closed than in the case of an open system.

The expression of $p(Y_1)$ for the open system, $p(Y_1)_0$, can be written as

$$p(Y_1)_0 = \frac{a+b}{c+d}$$ (9.25)

with:

$$a = \tau_2 \tau_3 \tau_4$$

$$b = \tau_0 \tau_2 \tau_4 \tag{9.26}$$

$$c = \tau_1 \tau_2 \tau_3 + \tau_1 \tau_2 \tau_4 + \tau_1 \tau_3 \tau_4 + \tau_2 \tau_3 \tau_4$$

$$d = \tau_0 \tau_2 \tau_4 + \tau_0 \tau_1 \tau_4$$

The expression of the same probability of occurrence, $p(Y_1)_C$, for a closed system, reduces to

$$p(Y_1)_C = \frac{a}{c} \tag{9.27}$$

It follows from these expressions that $p(Y_1)$ will be larger for an open than for a closed system if

$$bc > ad \tag{9.28}$$

This condition implies that

$$\tau_0 (\tau_1 \tau_2 \tau_3 + \tau_1 \tau_2 \tau_4 + \tau_1 \tau_3 \tau_4 + \tau_2 \tau_3 \tau_4) > \tau_3 (\tau_0 \tau_2 \tau_4 + \tau_0 \tau_1 \tau_4) \tag{9.29}$$

Simple algebraic calculation shows that this expression is equivalent to

$$\tau_3 + \tau_4 > 0 \tag{9.30}$$

which is, of necessity, fulfilled. It follows from this reasoning that

$$p(Y_1)_0 > p(Y_1)_C \tag{9.31}$$

The same reasoning can be applied to $p(Y_2)_0$ and $p(Y_2)_C$. The expression of $p(Y_2)_0$ is similar to that of $p(Y_1)_0$ with

$$a = \tau_1 \tau_2 \tau_4$$

$$b = \tau_0 \tau_1 \tau_4 \tag{9.32}$$

$$c = \tau_1 \tau_2 \tau_3 + \tau_1 \tau_2 \tau_4 + \tau_1 \tau_3 \tau_4 + \tau_2 \tau_3 \tau_4$$

$$d = \tau_0 \tau_1 \tau_4 + \tau_0 \tau_2 \tau_4$$

As previously, condition $bc > ad$ is fulfilled if $\tau_3 + \tau_4 > 0$. This reasoning allows one to conclude that

$$p(Y_1)_0 > p(Y_1)_C$$

$$p(Y_2)_0 > p(Y_2)_C \tag{9.33}$$

$$p(Y_3)_0 < p(Y_3)_C$$

$$p(Y_4)_0 < p(Y_4)_C$$

The conclusion that can be drawn from these results is that the open character of the network generates an increase of the probability of occurrence for certain nodes and a decrease of the probability of occurrence for others. Put in other words, this means that, in an open system, certain nodes take an importance they would not possess if the system were closed.

The conclusion that the node probabilities are dependant upon the fact that the network is open, or not, can be generalized to any type of metabolic network, or meta-network (Fig. **3**). The corresponding probabilities of occurrence of the nodes are

$$p(Y_1) = \frac{\tau_3 \tau_5 \tau_6 (\tau_4 + \tau_0)(\tau_2 + \tau_2')}{\Delta}$$

$$p(Y_2) = \frac{\tau_1 \tau_3 \tau_5 \tau_6 (\tau_4 + \tau_0)}{\Delta}$$

$$p(Y_3) = \frac{\tau_1 \tau_2 \tau_5 \tau_6 (\tau_4 + \tau_0) + \tau_1 \tau_2' \tau_4 \tau_5 \tau_6}{\Delta} \tag{9.34}$$

$$p(Y_4) = \frac{\tau_1 \tau_2' \tau_3 \tau_5 \tau_6}{\Delta}$$

$$p(Y_5) = \frac{\tau_1 \tau_2' \tau_3 \tau_4 \tau_6}{\Delta}$$

$$p(Y_6) = \frac{\tau_1 \tau_2' \tau_3 \tau_4 \tau_5}{\Delta}$$

with

$$\Delta = \tau_5 \tau_6 (\tau_4 + \tau_0)(\tau_1 \tau_3 + \tau_1 \tau_2 + \tau_2 \tau_3 + \tau_2' \tau_3) + \tau_1 \tau_2' (\tau_3 \tau_4 \tau_5 + \tau_3 \tau_4 \tau_6 + \tau_3 \tau_5 \tau_6 + \tau_4 \tau_5 \tau_6) \quad (9.32)$$

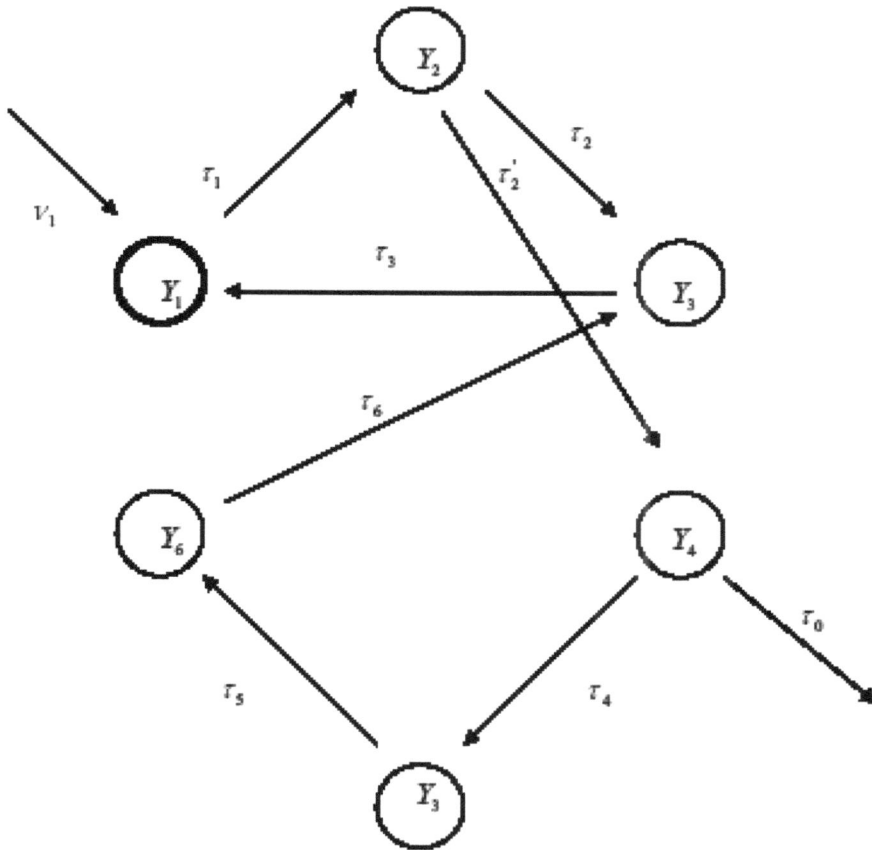

Figure 3: The graph description of a more complex open multienzyme system. As previously, each node is an enzyme reaction.

4- GENERAL CONCLUSIONS

Recently, a great deal of scientific papers has been devoted to networks [7, 8]. A basic idea underlying this literature is that they exist a *general* science of networks that would apply to any type of network [7, 8]. In this perspective, any kind of collective property could be described by a network. Owing to its generality, the unified science of networks could span the traditional boundaries between classical disciplines such as statistical physics, molecular, cellular, organismic and population biology, sociology, economics and other social sciences.

Such vision is far too optimistic for the properties of networks may not be independent from the nature of their nodes. For instance, the social relationships in a human community are not identical, and cannot be compared, to the node connexions in a biochemical network for a simple reason: biochemical networks should meet the laws of thermodynamics which is obviously not the case for social relationships in a human community.

In the case of a metabolic network, for instance, it is impossible to define node probabilities independently of the degree of node connexion. This situation is at variance with the concept of "hub", *viz.* a highly connected node with a poor probability of occurrence [7,8]. A network of social relationships, for instance, can perfectly be a closed, isolated, structure. This is not the case for metabolic networks that should be considered open dynamic structures in such a way that the probability of occurrence of any node is, in part, dependant upon the output of matter from the network. Node probabilities are in fact mathematical functions of the rates of connexion of the other nodes, including the output rate from the network.

In a metabolic network the free energy change upon going from a node to another one should be independent from the pathway. This implies the existence of some thermodynamic constraints when going from a node to another one through different pathways. Such constraints should not exist for networks that follow the laws of thermodynamics.

The concept of "distance" between nodes is essential for the definition of the so-called "small worlds" [8]. This concept of "distance" implicitly postulates that one can go from a node to another one and back following the same pathway, which is then considered fully reversible. In fact, many enzyme reactions are not fully reversible. Their degree of reversibility is limited by thermodynamic conditions. It follows from this remark that in order to go from a node, and back, one should follow different pathways that should possess different lengths. It might then follow that the concept of small world could not apply to metabolic networks

REFERENCES

[1]   Erdos, P. and Renyi, A. (1960) On the evolution of random graphs. Publi. Math. Instit. Hung. Acad. Sci. 5, 17-61.
[2]   Bolobas, B. (1985) Random Graphs. Academic Press, London.
[3]   Karp, P.D., Kummenaker, M., Paley, S. and Wagg, J. (1999) Integrated pathway-genome databases and their role in drug discovery. Trends in Biotechnology 17, 275-281.
[4]   Kanehisa, M. and Goto, S. (2000) KEGG: Kyoto encyclopedia of genes and genomes. Nucleic Acids Res. 28, 27-30.
[5]   Overbeck, R. *et al.* (2000) WITT: Integrated system for high-throughput genome sequence analysis and metabolic reconstruction. Nucleic Acid Res. 28, 123-125.
[6]   Jeong, H., Tombor, B.Albert, R. Oltvai, Z. N. and Barabasi, A. L. (2000) The large scale organization of metabolic networks. Nature 407, 651-654.
[7]   Albert, R. and Barabasi, A.L. (2002) Statistical mechanics of complex networks. Rev. Mod. Phys. 74, 47-97.
[8]   Barabasi, A.L. (2002) Linked: The new Science of Networks. Perseus Publishing, New York.
[9]   Barabasi, A.L. and Albert, R. (1999) Emergence of scaling in random networks. Science, 286, 509-512.
[10]   Albert, R. Jeong, H. and Barabasi, A. L. (2000) Error and attack tolerance of complex networks. Nature 406, 378-382.

CHAPTER 10

Non-Equilibrium Dynamics, Biological Systems and Time-Arrow

Abstract: The concept of time-arrow does not appear frequently in physics. This is not the case for the study of biochemical systems as many of them are able to sense whether a variable is increasing or decreasing. This effect can be explained by the non-linear character of the system and its dynamics.

Keywords: Bound enzymes and hysteresis loops, Time-arrow and non-linearity of sequences of biochemical reactions, Non-linearity and sensing chemical signals, Sensing the direction of a process.

As we shall see, non-equilibrium dynamic systems are often associated with some kind of time-arrow *viz.* they are able to sense whether the intensity of a signal increases or decreases. In a way, these dynamic systems mimic a fundamental property of living organisms which are able to sense whether a property is increasing or decreasing. In the perspective of biological systems theory it is important to know whether this important property can be understood and explained in physical terms.

1- BOUND ENZYMES AND HYSTERESIS LOOPS

It is relatively easy to build up an artificial system that is able to mimic a biological hysteresis loop. The equation that describes the system should, of necessity, possess a non-linear term in substrate concentration. Let us consider, as an example, an enzyme bound to an impermeable membrane as shown in Fig. **1** [1-6]. We assume that this enzyme is inhibited by an excess substrate. The corresponding reaction rate is then

$$v_e = \frac{VS(m)/K_1}{1 + S(m)/K_1 + K_2 S^2(m)/K_1} \tag{10.1}$$

In this equation, V is the maximum rate, $S(m)$ is the substrate concentration "seen" by the bound enzyme molecules, K_1 the apparent affinity constant of a substrate molecule for the enzyme and K_2 the equilibrium constant of substrate S for a site of the enzyme distinct from the active site.

Jacques Ricard

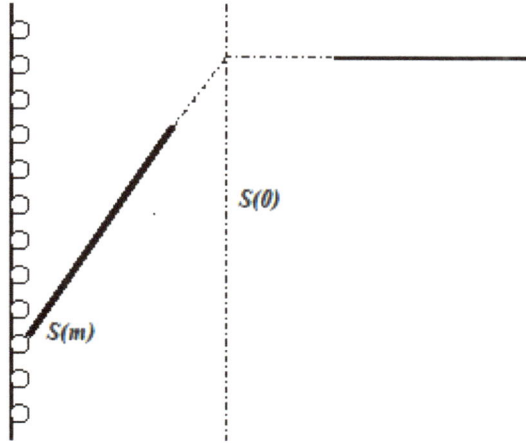

Figure 1: Gradient of substrate concentration in the vicinity of a membrane. See text.

If we assume that the substrate molecule S diffuses towards the bound enzyme one has

$$h_d \{S(0) - S(m)\} - \frac{VS(m)/K_1}{1 + S(m)/K_1 + K_2 S^2(m)/K_1} = 0 \qquad (10.2)$$

There is an obvious advantage in expressing this equation in dimensionless form. Setting

$$S_\sigma = \frac{S(m)}{K_1}$$

$$S_0 = \frac{S(0)}{K_1} \qquad (10.3)$$

$$h_d^* = \frac{h_d}{V}$$

$$\frac{K_2}{K_1} = \lambda$$

equation (10.2) becomes

$$h_d^*(S_0 - S_\sigma) - \frac{S_\sigma}{1 + S_\sigma + \lambda S_\sigma^2} = 0 \tag{10.4}$$

which is in fact a third degree equation in S_σ viz.

$$h_d^* \lambda S_\sigma^3 - h_d^*(\lambda S_0 - 1) S_\sigma^2 + \left\{h_d^*(1 - S_0) + 1\right\} S_\sigma - h_d^* S_0 = 0 \tag{10.5}$$

This relationship is in fact a coupling equation that associates a bound enzyme process with the diffusion of substrate in the vicinity of the membrane. According to the Descartes rule of signs, this third degree rate equation may possess three real positive roots. This implies that, when plotting S_σ as a function of S_0, the corresponding curve displays a S-shape, as shown in Fig. **2**. This means that the coupled system displays three steady states. Two of these steady states are stable and one is unstable. It follows from this situation that the variation of S_σ as a function of S_0 will follow different pathways depending on S_σ increases or decreases (Fig. **2**). The system is then able to sense not only the intensity of a signal but also whether this intensity is increasing, or decreasing. Such a situation, which allows one to compare this model to much more complex physiological situations, is due to the nonlinearity of the system and to the fact that the coupled system occurs under non-equilibrium conditions.

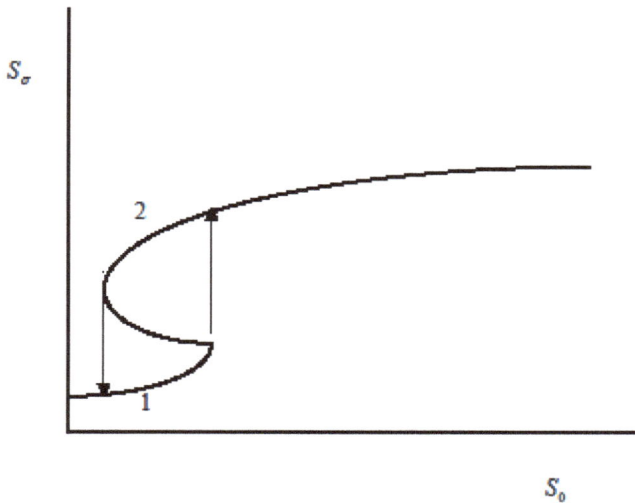

Figure 2: Hysteresis that generates time-arrow. In region 1 concentration can only increase. In region 2 it can only decrease.

2- NONLINEARITIES OF SEQUENCES OF BIOCHEMICAL REACTIONS AND TIME-ARROW

The simplest phenomenological description of an enzyme reaction is

$$E + S \underset{}{\overset{K}{\rightleftharpoons}} ES \overset{k}{\longrightarrow} E + S'$$

where S' is a product of the enzyme reaction. In this simple scheme, it is assumed that enzyme E binds substrate S to form an enzyme-substrate complex that decomposes as to regenerate the free enzyme and to give birth to the product S'. In this ideal scheme K is the ratio of rate constants involved in substrate binding and release followed by the desorption of the product S'. k is the apparent catalytic constant. If this ideal scheme is applied to a linear sequence of enzyme reactions one has

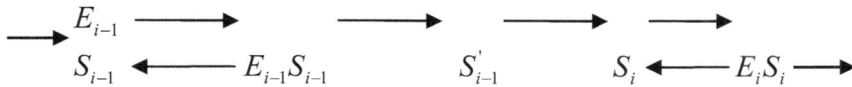

$$\longrightarrow \begin{matrix} E_{i-1} \longrightarrow \\ S_{i-1} \longleftarrow E_{i-1}S_{i-1} \end{matrix} \longrightarrow \begin{matrix} \\ S'_{i-1} \end{matrix} \longrightarrow \begin{matrix} \\ S_i \longleftarrow E_i S_i \end{matrix} \longrightarrow$$

It is assumed in this scheme that substance S_{i-1} binds to enzyme E_{i-1}, is transformed into S'_{i-1} that diffuses and binds to another enzyme E_i. As long as the substrate is in the vicinity of enzyme E_{i-1} its concentration is S'_{i-1}. When it comes in the vicinity of enzyme E_i its concentration is S_i. If diffusion is comparatively slow as compared to the catalytic process, the sequence above can be summarized as

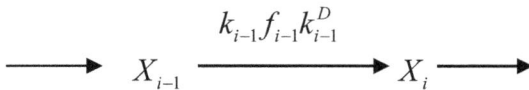

$$\longrightarrow X_{i-1} \overset{k_{i-1}f_{i-1}k_{i-1}^D}{\longrightarrow} X_i \longrightarrow$$

where

$$X_{i-1} = E_{i-1} + E_{i-1}S_{i-1} \tag{10.6}$$

and

$$X_i = E_i + E_i S_i \tag{10.7}$$

Moreover the expression of f_{i-1} is

$$f_{i-1} = \frac{E_{i-1}S_{i-1}}{E_{i-1} + E_{i-1}S_{i-1}} \tag{10.8}$$

This expression is a fractionation factor and can be considered as the probability, $p(S_{i-1})$, that substrate S_{i-1} binds to enzyme E_{i-1}.

We have previously mentioned that k_i^D is the diffusion constant of a reactant that moves from the immediate vicinity of enzyme E_{i-1} to enzyme E_i. The corresponding time constant, τ_{i-1}, of the transition from enzyme E_{i-1} to enzyme E_i is then

$$\tau_{i-1} = k_{i-1}k_i^D p(S_{i-1}) \tag{10.9}$$

The general idea expressed in this reasoning is that the transition from an enzyme reaction to another one involves the release of a reactant from an enzyme (E_{i-1}) and its transfer to another one (E_i). The rate of the corresponding diffusion process is then

$$v_i = k_{i-1}^D (S'_{i-1} - S_i) \tag{10.10}$$

Again it must be stressed that S'_{i-1} and S_i are different concentrations of the same substance and that $S'_{i-1} > S_i$. If now we assume, as done in the previous Section, that the enzyme reaction which uses S_i as a substrate is inhibited by an excess substrate, the corresponding enzyme reaction process is again described by equation (10.10). If now we assume, as done in the previous Section, that the enzyme reaction that uses S_i as a substrate is inhibited by an excess substrate, the corresponding enzyme reaction process is then

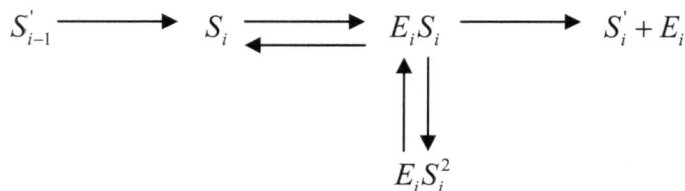

$$S'_{i-1} \longrightarrow S_i \;\rightleftarrows\; E_i S_i \longrightarrow S'_i + E_i$$
$$\big\updownarrow$$
$$E_i S_i^2$$

and the corresponding enzyme reaction rate can be represented as

$$v_i = \frac{V_i K_i S_i}{1 + K_i S_i + K_i K S_i^2} \tag{10.11}$$

In this expression, K_i is the binding constant of S_i to the free enzyme and K the binding constant of the S_i to the enzyme–substrate complex $E_i S_i$. V_i is the maximum reaction rate. In order to be able to write equations (10.10) and (10.11) in dimensionless form one can define the new dimensionless variables and parameters as

$$s_i = K_i S_i$$

$$s'_{i-1} = K_i S'_{i-1} \tag{10.12}$$

$$\lambda_K = \frac{K}{K'_i}$$

$$k_{i-1}^{D*} = \frac{k_{i-1}^D}{K_i V_i}$$

It is then possible to rewrite in dimensionless form the equations of diffusion and of the enzyme process. One finds

$$\frac{v_i}{V_i} = k_{i-1}^{D*}(S'_{i-1} - S_i) \tag{10.13}$$

and

$$\frac{v_i}{V_i} = \frac{s_i}{1 + s_i + \lambda_K s_i^2} \tag{10.14}$$

Hence, under steady state conditions, one has

$$k_{i-1}^{D*}(s'_{i-1} - s_i) - \frac{s_i}{1 + s_i + \lambda_K s_i^2} = 0 \tag{10.15}$$

This equation can be rearranged to

$$s_i^3 - \frac{\lambda_K s_{i-1}' - 1}{\lambda_K} s_i^2 + \frac{k_{i-1}^{D*}(1 - s_{i-1}') + 1}{k_{i-1}^{D*} \lambda_K} s_i - \frac{s_{i-1}'}{\lambda_K} = 0 \tag{10.16}$$

The important question that has to be worked out is to demonstrate that this equation can possess three positive real roots. In order to display such behavior this equation should be able to be rewritten as

$$s_i^3 - (\lambda_1 + \lambda_2 + \lambda_3)s_i^2 + (\lambda_1\lambda_2 + \lambda_1\lambda_3 + \lambda_2\lambda_3)s_i - \lambda_1\lambda_2\lambda_3 = 0 \tag{10.17}$$

where the roots $\lambda_1, \lambda_2, \lambda_3$ should be positive real numbers. If we compare equations (10.16) and (10.17) it appears that the coefficients λ_K, s_{i-1}' and k_{i-1}^{D*}, which are physical quantities, cannot possess negative values. Hence there should exist some constraints between these physical coefficients and the roots λ_1, λ_2 and λ_3. Comparing $\lambda_1 + \lambda_2 + \lambda_3$ and $\frac{\lambda_K s_{i-1}' - 1}{\lambda_K}$ leads to the conclusion that

$$\frac{\lambda_K s_{i-1}' - 1}{\lambda_K} = s_{i-1}' - \frac{1}{\lambda_K} = \lambda_1 + \lambda_2 + \lambda_3 \tag{10.18}$$

Moreover one should have

$$\frac{s_{i-1}'}{\lambda_K} = \lambda_1\lambda_2\lambda_3 \tag{10.19}$$

and therefore

$$s_{i-1}' = \lambda_K \lambda_1\lambda_2\lambda_3 \tag{10.20}$$

It then follows that

$$\frac{1}{\lambda_K} = \frac{\lambda_1\lambda_2\lambda_3}{\lambda_1 + \lambda_2 + \lambda_3} \lambda_K \tag{10.21}$$

and

$$\frac{1}{\lambda_K^2} = \frac{\lambda_1\lambda_2\lambda_3}{\lambda_1 + \lambda_2 + \lambda_3} \tag{10.22}$$

Hence one has

$$\lambda_K = \sqrt{\frac{\lambda_1 + \lambda_2 + \lambda_3}{\lambda_1 \lambda_2 \lambda_3}} \tag{10.23}$$

Combining expressions (10.20) and (10.23) leads to

$$s'_{i-1} = \sqrt{\lambda_1 + \lambda_2 + \lambda_3} \, \frac{\lambda_1 \lambda_2 \lambda_3}{\sqrt{\lambda_1 \lambda_2 \lambda_3}} = \sqrt{\lambda_1 + \lambda_2 + \lambda_3} \sqrt{\lambda_1 \lambda_2 \lambda_3} \tag{10.24}$$

It follows that the coefficient of the term in s_i of equation (10.16) can be expressed in terms of equation (10.17) and therefore possesses three positive real roots provided $1/\lambda_K < s'_{i-1} < 1$. Below the limit $s'_{i-1} = 1/\lambda_K$ and above $s'_{i-1} = 1$, equation (10.16) has one real and two imaginary roots. This situation is depicted in Fig. **3**.

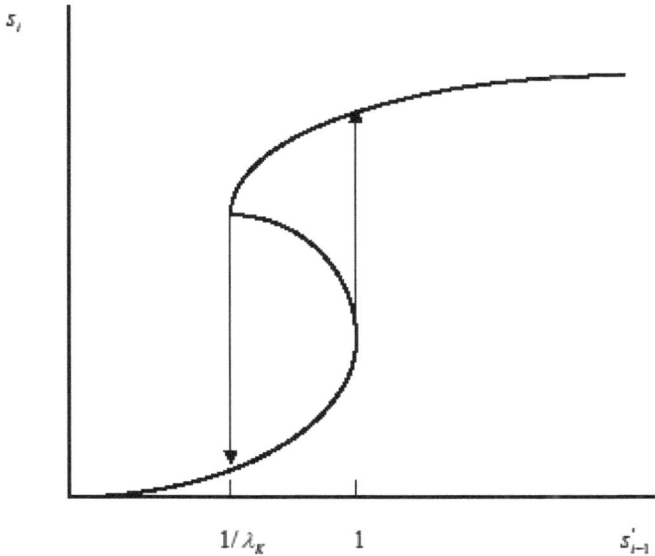

Figure 3: Hysteresis and time-arrow for a sequence of enzyme reactions. See text.

In the interval $1/\lambda_K < s'_{i-1} < 1$ the system displays some kind of chemical hysteresis and depending on s'_{i-1} increases or decreases, the system will follow different routes. The situation depicted in Fig. **3** mimics the fact that living systems are perfectly able to sense whether the intensity of a signal increases, or decreases, and are able to react accordingly. Such a situation is due to a non-linearity in equation (10.15).

3- DEPENDENCE OF THE ENZYME SYSTEM UPON ITS HISTORY

If we neglect product inhibition, the progress curve of product appearance during an enzyme reaction can be expressed as

$$\frac{[P]}{[E]_T} = \alpha + \frac{v_s t}{[E]_T} + \sum_{i=1}^{n} \psi_i \exp(-\lambda_i t) \tag{10.25}$$

where $[P]$ is the concentration of the product at time t, $[E]_T$ the total enzyme concentration, v_S the steady state velocity, t the time, α, ψ_i and λ_i groupings of rate constants and ligand concentrations. Classical enzyme kinetics implicitly, or explicitly, considers that, as time t increases, the exponential terms in equation (10.25) vanish and the progress curve is, for a long period of time, linear. However, if the enzyme undergoes a slow conformation change, one has

$$\exp(-\lambda_n t) \approx 1 - \lambda_n t \tag{10.26}$$

Under these conditions equation (10.25) becomes

$$\frac{[P]}{[E]_T} = \alpha + \psi_n + \left(\frac{v_s}{[E]_T} - \lambda_n \psi_n \right) t + \sum_{i=1}^{n-1} \psi_i \exp(-\lambda_i t) \tag{10.27}$$

Hence the steady state that can be measured is only apparent and is called, for that reason, meta-steady state. Its expression is

$$v_S^* = v_S - \lambda_n \psi_n [E]_T \tag{10.28}$$

Whereas the true steady state of the system does not depend upon the history of the reaction, the meta-steady state does for ψ_n is dependent upon the *direction* of the process. If, for instance, we follow the steady state kinetics of an enzyme reaction upon increasing, or decreasing, the concentration of a substrate, or that of any other ligand, involved in this reaction, one usually expects that the rate of the process be the same for the same ligand (or substrate) concentration whether this concentration is reached after an increase, or a decrease. This is not the case for the enzyme fructose bisphosphatase [7]. If the enzyme is incubated in the presence of the substrate analogue fructose 2, 6 bisphosphate, it becomes activated. However, if the enzyme activity is measured as a function of fructose bisphosphate concentration, the reaction rate is different depending on a given concentration is reached after an increase, or a decrease (Fig. **4**). This means that the enzyme activity is dependent upon both a ligand concentration and the history of the system.

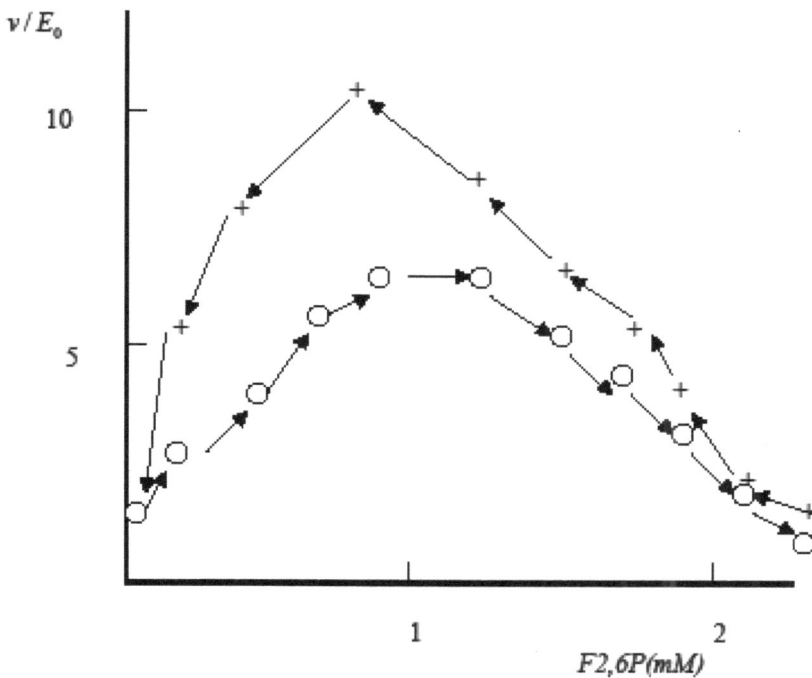

Figure 4: Time-arrow in the activation-deactivation of an enzyme. See text.

A mechanistic device that may generate this kind of behavior is the so-called "mnemonical model" [8]. In its simplest version, the free enzyme exists under two conformations (the circle and the rhombus, Fig. **5**). If the rhombus has a stronger affinity for substrate S than does the circle, the enzyme process $S \rightarrow P$ will be slow, for part of the enzyme (the circle) will not be used in this process. However, the reverse process $P \rightarrow S$ will be faster for most of the enzyme will be in the rhombus conformation that binds strongly substrate S.

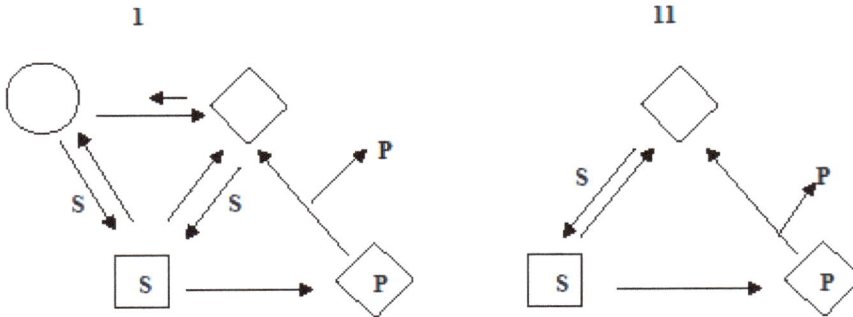

Figure 5: The "mnemonical model" as a possible explanation of time-arrow. See text.

4- GENERAL CONCLUSIONS

A fundamental property of living systems is their ability to sense whether the intensity of a signal is reached after an increase, or a decrease, of intensity. These systems are then sensitive to some kind of a time-arrow. This is interesting for most simple physical systems do not possess this property. Thus for instance the fundamental relation of dynamics $F = m\partial^2 x / \partial t^2$ remains unchanged whether t increases or decreases. Physicists have often discussed the question of the reality of time-arrow for simple physical systems [9]. With the notable exception of Prigogine, most of them seem to be convinced that such a time arrow does not exist. There is little doubt, however, that simple biological systems possess this property and are able to react differently depending on t (the time) is increasing or decreasing. An interesting idea that has been proposed in modern non-equilibrium thermodynamics is that the ability to react to time-arrow is linked to the non-linearity of the system. This is particularly evident with equation (10.16). If this equation were linear in s_i one could not have expected multiple steady states to occur and hence different types of behavior of the system depending on the variable s_i reaches the same value after an increase, or a decrease.

These considerations may be of value in the general context of the origins of life. If we imagine the minimum conditions required for a network of chemical reactions to display some of the features of "simple" living systems there is little doubt that one of these conditions is to display non-linearities as to exhibit multiple steady states. Indeed, these conditions are far from being sufficient but they are no doubt necessary as to display some of the main features of living systems.

REFERENCES

[1]   Engasser, J. M. and Horvath, C. (1974a) Inhibition of bound enzymes. I. Antienergistic interaction of bound chemicals and diffusional inhibition. Biochemistry 13, 3845-3849.

[2]   Engasser, J. M. and Horvath, C. (1974b) Inhibition of bound enzymes. II. Characterization of product inhibition and accumulation. Biochemistry 13, 3849-3854.

[3]   Engasser, J. M. and Horvath, C. (1974c) Inhibition of bound enzymes. III Diffusion enhanced regulatory effect with substrate inhibition. Biochemistry 13, 3855-3859.

[4]   Thomas, D., Barbotin, J. N.,Hervagault, J. F. and Romette, J. L. (1977) Experimental evidence for a kinetic and electrochemical memory in enzyme membranes. Proc. Natl. Acad. Sci. USA 74, 5314-5317.

[5]   Ricard, J. and Noat, G. (1984a) Enzyme reactions at the surface of living cells.I. Electric repulsion of charged ligands and recognition of signals from the charged milieu. J. Theor. Biol. 109, 555-569.

[6]   Ricard, J. and Noat, G. (1984b) Enzyme reactions at the surface of living cells. II. Destabilization in the membrane and conduction of signals. J. Theor. Biol. 109, 571-580.

[7]   Soulié, J.M., Rivière, M., Baldet, P. and Ricard, J. (1991) Kinetics of the conformational transition of the spinach chloroplast fructose 1,6-bisphosphatase. Eur. J. Biochem. 195, 671-678.

[8]   Ricard, J., Meunier, J. C. and Buc, J. (1974) Regulatory behavior of monomeric enzymes. I. The mnemonical enzyme concept. Eur. J. Biochem. 49, 195-208.

[9]   Gell-Mann, M. (1994) The quark and the jaguar. Adventures in the simple and the complex. French translation: Le quark et le jaguar. Voyage au Coeur du simple et du complexe. Albin Michel, 1995 Paris.

Send Orders for Reprints to reprints@benthamscience.net

CHAPTER 11

Some Philosophical Implications of the Concept of Biological System

Abstract: The present Chapter relates and summarizes the main features of biological systems. Contrary to molecular biology, systems biology does not attempt at explaining biological events in terms of the properties of biomolecules but in terms of *systems* of biomolecules. As these systems are abstract entities it follows that systems biology has an abstract character that can only be approached by the use of mathematics. Exactly as for physics, this situation raises the question of the concrete reality of the theoretical concepts.

Keywords: Systems and emergent properties, Systems and information, Systems and Plato's metaphysics, Systems and mathematics, Reality and objectivity of mathematical concepts, Systems and artificial life.

The new science of systems biology constitutes a somewhat novel approach of biological problems. Instead of studying the functional properties of living organisms through the structure and function of some of their macromolecules, such as nucleic acids and proteins, systems biology attempts at considering a living organism, or part of a living organism, as a system involving many *relations* between biomolecules. In this perspective, biological properties are not defined at the molecular level but at the level of a multi-molecular system.

1- SYSTEMS BIOLOGY AND MOLECULAR BIOLOGY

As it has appeared in this book, the aim of classical molecular biology is to explain biological phenomena from a precise knowledge of the structure and function of biological macromolecules. This reasoning implies that biological properties can be reduced to the properties of a number of biomolecules such as nucleic acids and proteins. The aim of systems biology is different. This new science considers that biological processes are not necessarily borne by macromolecules but by *systems* of macromolecules that mutually interact. In this perspective, the biologist does not necessarily accept the view that a biological property is borne by a macromolecule but rather by a *system* of interacting macromolecules. The behavior of the system is then to be explained not necessarily by the structure of these macromolecules but by the functional interactions they can mutually display.

2- THE ABSTRACT CHARACTER OF THE CONCEPT OF SYSTEM

Such a set of macromolecules that mutually interact is more than a simple set. It is a *system* that possesses properties different from those of individual macromolecules. The system is in fact a somewhat abstract entity whose properties emerge from the interactions occurring between different macromolecules. In this perspective, it is obvious that the interactions between macromolecules, or between macromolecules and charged ligands, may generate properties that are different from those of the same macromolecules considered in isolation.

3- EMERGENT PROPERTIES OF SYSTEMS

As previously mentioned, the global properties of systems are emergent with respect to the individual properties of their constituents and hence cannot be reduced to them. The logic of these global properties cannot be understood on a simple intuitive basis but can easily be approached through the use of mathematics. In fact, novel properties can emerge out of the interactions that may occur between the elements of such systems and are therefore predictable through the use of mathematics. An important idea of systems biology is the importance of the mathematical tool for understanding the organization and the functioning of such systems.

4- SYSTEMS AND INFORMATION

In classical molecular biology, information is, in general, borne by DNA and transferred to RNA, then to proteins. It cannot be generated *de novo* but only through the duplication of already existing DNA molecules. We have seen previously that, in fact, new information can be generated in a system provided this system is not submitted to the subadditivity principle. Such a situation, for instance, may possibly occur if two ligands, x and y, bind to a protein and mutually interact. If the resulting information, $h(x|y)$ is larger than $h(x)$ then, under these conditions, the system *generates* its own information.

5- SYSTEMS AND MATHEMATICS

The use of mathematics is compulsory in systems biology because the important point, in a system, is to study the *relations* between material entities. These relations may not be *material* but even so they are *real* for they control the behavior of the system. We are then led to a very old problem of philosophy, the problem of *ideas* and *universals* that dates back to Plato [1] and Aristotle [2] and has been discussed by many philosophers such as Russell [3, 4]. The way this problem is discussed by Plato is the following. Plato, for instance, stresses the point there exists virtuous acts but these virtuous acts are different from virtue *per se*. Virtue implies some kind of *relation* between many virtuous acts. According to Aristotle, a relation is somewhat different from physical objects and sense data, but are related to mathematics as mathematical entities, equations for instance, relate something to something else. In this perspective, the mathematical description of a system should be considered real and this leads us to the well known problem of the nature of laws, as expressed by mathematics. It is noteworthy that most experimentalists who use mathematics as a tool do not consider equations as real entities independent of the mathematician. Implicitly, or explicitly, they think that mathematical objects have been *invented* by mathematicians and would not exist in their absence.

Mathematicians, however, are strongly convinced they have not *invented* anything. They have *discovered* mathematical entities and theorems that exist in an immaterial, but *real world*. In fact, many important mathematical discoveries contradict common sense but have to be accepted as the inescapable implications of valid theorems. Such a situation has been occurring very early in the history of mathematics with the irrational character of $\sqrt{2}$, the complex numbers, sets theory, *etc...* Systems biology is, to some extent, in a situation that is somewhat similar to theoretical physics where the main results can in fact be predicted by a theory. The reality of theoretical concepts have been strongly advocated by philosophers of Science, such as K. Popper [5], and physicists, such as Penrose [6] who believe that, in addition to the physical and the mental worlds, there exists a world of concepts and laws.

6- SYSTEMS, COMPUTATION AND EVOLUTION

Living systems are subjected to evolution and this evolution can be mimicked in artificial systems. We have seen, for instance (Chapter 4), that enzyme reactions may, or may not, display emergence of information, depending on the values of their rate constants. Slight random perturbations of the medium may alter the rate properties of the system in such a way that its global behavior may be altered. Such a situation can be shown, by computer simulation, to take place. The systems that are best fitted to the external milieu will be selected whereas the others will disappear.

The computational study of evolution of a system allows to study the emergence of novel "biological" properties. For that reason it can be considered some kind of "experimental" approach of evolutionary processes.

7- SYSTEMS AND ARTIFICIAL LIFE

The use of mathematics to study biological problems has made possible the approach of an unexpected question: the question of artificial life. In classical molecular biology this question does not make sense for it is, implicitly or explicitly, considered that life has a special status and cannot be engineered by man. There are several reasons that apparently justify this belief. One of these reasons originates from molecular biology itself. According to the so-called "central dogma" of this science, all the information of a living organism is initially stored in DNA and transferred to RNA and proteins, in succession. If this were always true, and had always been true in the past, it would be extremely difficult to offer a sensible scenario of the origins of life on earth. As a matter of fact, in this scenario, DNA is necessary for protein synthesis but proteins are necessary for DNA synthesis…. In order to circumvent this difficulty, the existence of a primordial RNA world has been postulated [7-9]. In this primordial world RNAs were supposed to replicate and to catalyze chemical reactions. In fact, if some RNAs have been isolated that are able to catalyze chemical reactions, these RNAs were unable to replicate in the absence of proteins [7-9]. For these reasons, some biologists have proposed that the prebiotic systems could have been networks of chemical reactions [10]. But then another problem appears. In such a system two important events should simultaneously emerge, namely its identity and its information. In classical Shannon communication theory there cannot exist any generation of information but only communication from place to place of a pre-existing information. In order to explain emergence of both identity and information, one has to have recourse to a chemical network that generates both its information and identity. As it has been previously discussed, network intermediates can possibly bind two different types of ligands x_i and y_j in such a way that

$$h(x_i|y_j) > h(x_i) \tag{12.1}$$

If such a situation takes place, we have seen (Chapter 3) there is spontaneous emergence of information in the system. This process is associated with a sequence of precise values of $h(x), h(y), h(x|y)$ and it is $h(y|x)$ that defines the identity of the biochemical network. Such considerations allow the biochemical network to possess an identity and to generate the information required for the occurrence of a given "biological" process.

8- GENERAL CONCLUSIONS

Systems biology puts the emphasis on the idea that many biological processes are not borne by macromolecules but rather by a system originating from the interactions that exist between these macromolecules. These interactions may be subtle and can only be described by mathematical models. In this perspective, biological functions themselves are also described by mathematical functions that connect molecular entities, or groups of molecular entities.

In other words, the logic of the system which is the support of a biological function is an abstract entity that connects different material entities. In this spirit, biological functions are described by equations that can be considered, in an Aristotelian perspective, a *universal*. Even though these equations are abstract entities they are nevertheless "real" and give the system both its identity and information.

REFERENCES

[1]   Platon, Parménide ou des Idées, Oeuvres completes, Bibliothèque de la Pléiade, tome II, pp.192-253.

[2]   Russell, B. (1995) History of Western Philosophy. Routleddge, London.

[3]   Russell, B. (1999) The Problems of Philosophy. Dover Publications, Inc. New York.

[4]   Aristote, Métaphysique, Tomes 1 et 2. Librairie Philosophique J. Vrin, Paris.

[5]   Popper, K. R. (1959) The logic of scientific discovery, Hutchinson, London.

[6]   Penrose, R. (2005) The Road to Reality. Vintage Books, London.

[7]   Cech, T. R. (1986) A model for the RNA-catalyzed replication of RNA. Proc. Natl. Acad. Sci. USA 83, 4360-4363.

[8]   Doudna, J. A.and Szostak, J. W. (1989) RNA catalyzed synthesis of complementary strand RNA. Nature, 339, 519-522.

[9]   Cech, T. R., Herschlag, D.,Piccirilli, J. A and Pyle, A.M. (1992) RNA catalysis by a group I ribozyme: developing a model for transition state stabilization. J. Biol. Chem., 267, 17479-17482.

[10] Kauffman, S. (1993) The Origins of Order; Oxford University Press.

Subject Index

A

Actin	127
Active transport	104
Affinity of a reaction	100
Allosteric model	6, 42
Apparent co-operativity of polyelectrolyte-bound enzymes	89

B

Biochemical networks as "open" systems	175
Biological systems	13

C

Central Dogma	21
Cis side and *trans* side	113
Communication channel	20, 29
Communication theory	19
Compartments	101
Complexity of polyelectrolyte matrices	81
Concentration control coefficients	70
Conditional information	17, 30
Control of differenciation	10
Co-operativity	42
Coupling of scalar and vectorial processes	69

D

De Donder inequality	100
Disfavoured reactions and cell compartments	104
DNA message	7, 18
DNA	7
Donnan equation	82
Dynamics of coupled reactions	69

E

Elasticity	73

www.ingramcontent.com/pod-product-compliance
Lightning Source LLC
Chambersburg PA
CBHW041658210326
41598CB00007B/455